What we know we know.

What we know we don't know.

What we don't know we don't know.

Given that we are only aware of

2,000 BITS

of information
out of the

400 BILLION BITS

of information
we are processing
per second . . .

When we argue against new knowledge . . .
how much of our "awareness" is arguing?

How can we know everything about
all the things we don't know?

What tHe βLEEP Ďθ wΣ (k)πow!? ™

DISCOVERING THE ENDLESS POSSIBILITIES FOR
ALTERING YOUR EVERYDAY REALITY

WILLIAM ARNTZ

BETSY CHASSE

MARK VICENTE

Health Communications, Inc.
Deerfield Beach, Florida

www.hcibooks.com
www.whatthebleep.com

Library of Congress Cataloging-in-Publication Data
is available from the Library of Congress

Publisher: Health Communications, Inc.
            3201 S.W. 15th Street
            Deerfield Beach, FL 33442-8190

Cover and inside book design by Larissa Hise Henoch
Inside book formatting by Lawna Patterson Oldfield
Original artwork for Chapter 11 by Gloria Naylor
Alchemy images in chapter 12 copyright ©Adam McLean 2002

CONTENTS

A Few Words from the Authors . . .

We started this project like we knew. We ended up, well, you know the name of the movie . . .

So if you're standing in the aisle of the bookstore reading this, and you're looking for some guide to tell you "how to do it" from way advanced beings who have already "done it," you probably want to put this book back on the shelf.

But if you are still reading, we're going on a trip.

Or rather, we went on a trip. We trooped around the country interviewing all these brilliant people so we could get on film what they had to say. Turns out we wanted to get on film what we thought they had to say and soon learned that what they really did have to say was *different*. Different from some of our notions, different from each other, different from what we were taught in school, different from what was preached in church and different from what we see on nightly news. And in the end we were the ones to decide. To decide *for ourselves* where the truth lies and what to try out in our lives.

Will Arntz (left) and world renowned scientist Dr. Ervin Laszlo

There seems to be a human tendency to think there's the magic formula, secret esoteric technique or hidden tradition that will suddenly make it all work. If there is some easy formula like this, in our combined decades of spiritual practice, we haven't found it.

So since we don't seem to know that much, you may wonder what we're doing writing a book. Well, for many people, *What the*

BLEEP introduced many new ideas and ways of looking at the world. For others it seemed to validate what they've always *felt* to be true, but never knew anyone else thought that way. So part of the book is informational: digging deeper into the science. Then there are the parts that reflect back on ourselves, how we perceive (or don't), what we do, and how our attitudes impinge upon our experience and reality. We also explore what some researchers have found that may give a key into why we do what we do.

And then there's the crystal ball. Many of the people we interviewed are visionaries, pioneers and prophets. We all sense we're on the verge of *something*; something big. Throughout history major shifts in the way people view the world (i.e., paradigms) are preshadowed by visionaries who either felt it coming and/or led the way. Did the visionaries create the new paradigm, or did the new paradigm go back in time and create the visionaries? Or does one thing create another, or, as some new models suggest, there is no one creating another, but only a state of mutual existence where cause and effect is replaced by *is*?

That's a rabbit hole. And as mystical as it sounds and bizarre as it reads, there *is* some scientific data that suggests it is indeed so.

We found interviewing our Greek Chorus of scientists, philosophers and mystics to be completely engaging. So much so that by the end of the interview, our film crews usually ended up jumping in and asking questions. And these crews were not people familiar with the material. They were film professionals we picked up in each city. And when they got exposed to these ideas and concepts, they couldn't help but get jazzed and start to wonder about the possibilities. And *that's* why we're writing this book—because people, lots of people, are interested in these subjects. And many of those interested didn't know they were until they got a taste. So if we can serve up some "quantum cooking," we are happy to do so. Have fun with it all, because it's a real trip into some really amazing stuff.

—*Will Arntz*

Five years ago I remember complaining to myself (as I often did) that Hollywood just wasn't making the kinds of films that I really felt were worthwhile and the world needed. I was a cinematographer at the time, and what I really wanted was to find directors to work with who wanted to make transformative films. All the while, I kept complaining about the film industry being shallow. One day, stark reality hit me, and I realized that it wasn't Hollywood's job to make life-altering films. Maybe it was mine. And all these years I've been blaming "it" for not meeting my requirements. Rather arrogant, wouldn't you say?

It never occurred to me that instead of complaining about it, I should just start directing—duh! Shortly after that I met William Arntz, who is probably one of the bravest men I know. Very, very few people have put their money where their mouth is, the way Will has. Together with the very talented Betsy Chasse, we formed a creative partnership that would give birth to this film, this book and a new sense of self for each of us. For the three years it took to make the film, we lived all the emotional addictions you see in the film, and have come out the other end older and wiser. This book contains a few more thoughts that never quite made it into the film, but also the same concepts and same information that made the film such a hit. I believe this knowledge and information is life-altering stuff. Enjoy our glimpse into the future of humanity.

—*Mark Vicente*

Four years ago I was living happily in my shoe consciousness. (What kind of shoes do I wear? What kind of car am I driving?) Then this film literally landed in my lap. Talk about the universe sending you a message! I have spent most of the last four years asking everyone and anyone, "What does this have to do with me?" and "How do I use this in my

life?" Trying to even begin to understand most of this stuff and then incorporate it can sometimes feel overwhelming. This book has been an amazing opportunity to explain our understanding of the wacky, weird world we live in. I hope that it helps you begin to do the same. This is our journey, our experiences and our take on it all. I don't profess to be a teacher or guru—but I can say that having the experience of making it into a film and then writing it all down in this book has changed me forever. I hope you'll find some useful info. But don't take our word for it—try it out yourself.

—*Betsy Chasse*

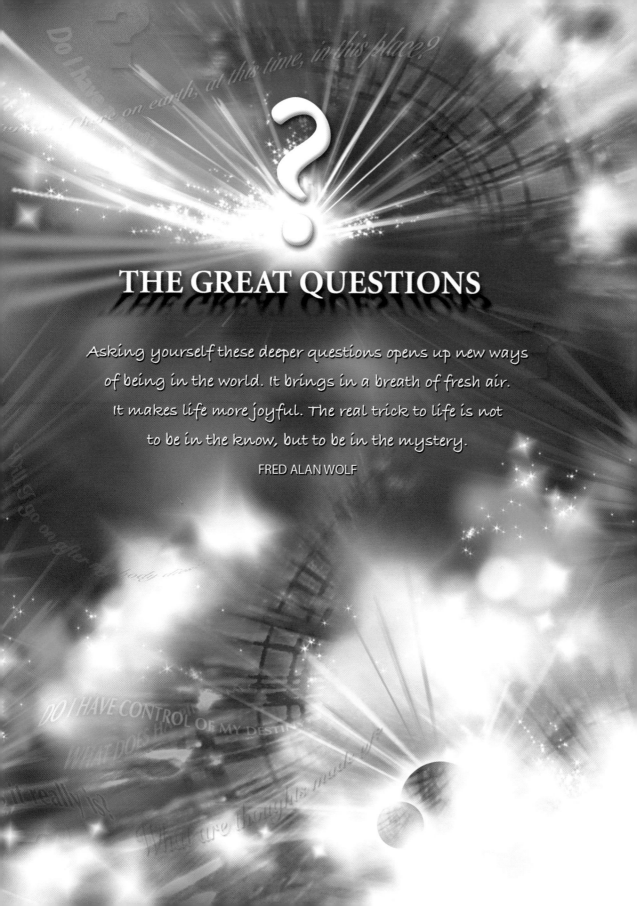

THE GREAT QUESTIONS

Asking yourself these deeper questions opens up new ways
of being in the world. It brings in a breath of fresh air.
It makes life more joyful. The real trick to life is not
to be in the know, but to be in the mystery.

FRED ALAN WOLF

What IS a Great Question?
Why should we bother?
What makes it Great?

Let's say a spaceship lands next to you on the coffee table (does size matter?) and inside is *The Universal Book of Everything*. And you get to ask one question. What is that question?

This may seem a little silly, but it's worth the effort. Take a minute and think. What would that question be? It can be anything. Go ahead and write it down in a journal.

Now let's say The Book is feeling a little underutilized these days, and you get a bonus question. Think of something that you are just plain curious about. It can be wondering if Elvis is still alive, or where you left your car keys. Something that simply tickles your fancy. Write that down, too.

And by now The Book is feeling a little depleted, and it got to be *The Universal Book of Everything* by asking questions of everyone and getting real answers. So, the question for you (the answer to which will be added in The Book) is:

What is the One Thing you know for sure?

Great Questions—The Can Opener of Consciousness

Aside from the few like Fred Alan Wolf (who we quoted on the opening page), when do we ever get encouraged to ask questions? And yet, most of those grand discoveries and revelations that our society cherishes came from asking questions. Those things, those *answers,* that we study in school came from *questions.* Questions are the precursor, or first cause, in every branch of human knowledge. The Indian sage Ramana Maharshi told his students the path to Enlightenment was summed up in: "Who am I?" The physicist Niels Bohr asked, "How can an electron move from A to B, and never go in between?"

These questions open us up to what we previously didn't know. And they're really the only way to get there—to the other side of the unknown.

Why ask a Great Question? Asking a Great Question is an invitation to an adventure, a journey of discovery. It's thrilling to set out on a new adventure; there's the bliss of freedom, the freedom to explore new territory.

So why don't we ask these questions? Because asking questions opens the door to chaos, to the unknown and unpredictable. The minute you ask a question you truly don't know the answer to, you open yourself up to a field of all possibilities. Are you willing to receive an answer you may not like or agree with? What if it makes you uncomfortable, or carries you outside the zone of safety and security you've built for yourself? What if the answer isn't what you want to hear!?

It doesn't take muscles; it takes bravery to ask a question.

Now let's consider what makes a question Great. A Great Question doesn't have to come from a philosophy book, or be about Life's Big Issues. A Great Question for you might be, "What would happen if I decided to go back to college and get a degree in a new field?" or "Should I listen to that voice that keeps telling me to go to California or China?" or "Is it possible to discover what is inside a neutrino?" Asking any of these questions and thousands of others could change the direction of your life. That's what a Great Question is: one that can change the direction of your life.

So, once again, why don't we ask them? Most people would rather stay in the safety of the known than go looking for trouble. Even if they crash right into a question, more than

Where do we come from?
What should we do?
And where are we going?

—Miceal Ledwith

The difference between me at five and me now is that at five I didn't have much invested emotionally in the Universe being a certain way. Being "wrong" never was a concern. It was all learning. Now I keep reminding myself: In science there is no such thing as a failed experiment. Learning that what I was testing simply does not work is actually a success.

—WILL

likely they will run away from it, stick their head in the sand or quickly get busy doing something else.

For most of us, it takes a serious crisis to bring on the Great Questions: a life-threatening illness, the death of someone close, failure of a business or a marriage, a repeated, even addictive behavior pattern you just can't seem to shake, or loneliness that seems unendurable for one more day. At times like those, Great Questions come boiling up from the depths of our being like hot lava. These questions are not intellectual exercises, but cries of the soul. "Why me? Why him? What have I done wrong? After this, is life truly worth living? How could God allow this to happen?"

If we could muster up the same kind of passion to ask ourselves a Great Question about our lives *now,* when there is no immediate crisis, who knows what could happen?

As Dr. Wolf said, asking a Great Question can open up new ways of being in the world. It can be a catalyst for transformation. Growing. Outgrowing. Moving on.

The Joy of Questions

Remember when you were five years old and you kept asking, "Why?" Your parents may have thought, after a while, that you were doing it simply to drive them crazy, but you really wanted to know! What happened to that five-year-old?

Can you remember the five-year-old who was you? Can you feel what it was like? This is important, because when you were five, you loved being in the mystery. You loved wanting to figure things out. You loved the journey. Each day was filled with new discoveries and new questions.

So what is the difference between then and now?

Good question!

The fun and joy of life are in the journey. In our culture, we've been conditioned to look at "not knowing" as something unacceptable and bad; it's some kind of failure. In order to pass

When I watch my daughter playing with whatever toy or gadget she has at the moment, I can see the pure joy in her face as she's trying to figure it out. She doesn't get discouraged; she just keeps on trying until she gets what she wants. Once she gets it, it's on to the next challenge, the next question.

I watched her this morning trying to figure out how to open the latch on the cupboard. It took a while, but she kept at it until she worked it out. Once she got the latch, the next joy was: Let's open the door! As the door opened, her face lit up with excitement. Look what's inside! What's that? There on the shelf? It was a true journey of discovery, full of joy at every step.

What I asked myself, and what I would ask you, is: What's *your* latch? What do *you* desire to know today?

—BETSY

the test, we have to know the answers. But even when it comes to factual knowledge about concrete things, what science *doesn't* know far exceeds what it *does*. Many of the greatest scientists have looked into the mystery of the universe and of life on our planet, and have frankly said, "We know very little. Mostly we have a lot of questions." This is certainly true of the outstanding thinkers we interviewed. In the words of author Terence McKenna, "As the bonfires of knowledge grow brighter, the more the darkness is revealed to our startled eyes."

It's even more difficult to come up with a clear-cut answer to "What is the meaning and purpose of my life?" The answer to Great Questions like this can only emerge from the journey of living. And we can only arrive at it by the road of not knowing— or maybe we should say, not-yet-knowing. If we always think we know the answer, how will we grow? What will we ever be open to learn?

> A university professor visited Zen master Nan-in to inquire about Zen. But instead of listening to the master, the scholar kept going on and on about his own ideas.
>
> After listening for some time, Nan-in served tea. He poured his visitor's cup full, and then kept on pouring. The tea flowed over the sides of the cup, filled the saucer, spilled onto the man's pants and onto the floor.
>
> "Don't you see that the cup is full?" the professor exploded. "You can't get any more in!"
>
> "Just so," replied Nan-in calmly. "And like this cup, you are full of your own ideas and opinions. How can I show you Zen unless you first empty your cup?"

Emptying the cup means making room for Great Questions. It means being open, reconditioning ourselves so that we can accept, for the time being, *not* knowing. Out of that a greater knowing will dawn.

I have found that I find a particular excitement in suddenly realizing I don't know the answer to something. It's like coming to the edge of a cliff in my mind.

In this space of "nothing" or not knowing, I get this intense feeling of anticipation. The reason I get excited is because I've come to the edge of what I know, and I realize that shortly an understanding will arrive in my head that will be staggering and will not have existed in me the moment before.

It will be this huge ah-ha. I learned recently that an ah-ha stimulates the pleasure center of the brain . . . Evidently I'm addicted to this feeling.

—MARK

Every age, every generation has its built-in assumptions— that the world is flat, that the world is round. There are hundreds of hidden assumptions, things we take for granted that may or may not be true. In the vast majority of cases, these conceptions about reality— which belong to the prevailing paradigm or worldview— aren't accurate. So if history's any guide, much that we take for granted about the world today simply isn't true.

—John Hagelin, Ph.D.

IT'S OK NOT TO KNOW THE ANSWER

A little while ago my sixteen-year-old niece sent me a long email. The gist of it was, "Life sucks. I see my dad coming home from work every day totally bummed out. I don't want to get trapped in the rat race, but I don't see any hope of avoiding it. Is this what life is about? What's the point? I might as well just shoot myself and die."

"Christina," I wrote back, "you might not think this is a great response, but I'm proud of you. I can't tell you that you are going to solve your dilemma and find The Answer. I know you want answers—but sometimes life doesn't provide them right away. But you are asking the right questions, and that is important."

—WILL

You're in Distinguished Company

People have been asking Great Questions for thousands of years. There have always been men and women who gazed at the stars and wondered at the vast mystery of it all, or who looked at the way people around them were living and thought, "Isn't there more to life than this?"

The ancient Greek philosophers pondered and discussed the Great Questions. Some, like Socrates and Plato, asked, "What is Beauty? What is Goodness? What is Justice? What is the best way to govern a society? What people are fit to be rulers?"

Religious teachers, mystics and spiritual masters like Buddha, Lao Tse, Jesus, Muhammad, St. Francis, Meister Eckhardt, Apollonius of Tyana and many more, in all the world's traditions, have asked Great Questions.

People with scientific minds have always asked questions. How does it work? What's inside? Are things really the way they seem? Where does the universe come from? Is the Earth the center of the solar system? Are there laws and patterns that underlie what happens in daily life? What's the connection between my body and my mind?

For the great scientists of history, these questions elicit a passion to *understand* that goes way beyond curiosity. They're not just curious—they need to know!

When Albert Einstein was a boy, he asked himself: "What happens if I'm riding my bicycle at the speed of light and I switch on my bike light—will it come on?" He nearly drove himself crazy asking himself that for ten years, but out of that resolute pursuit came the relativity theory. This is a great example of asking a question and hanging with it for years, in the unknown, until he came up with a completely different view of reality.

Paradigm Busting

One of the great things about science is its assumption that what it thinks it knows today will probably be proven wrong tomorrow. The theories of yesterday have served as platforms to climb higher, as Sir Isaac Newton meant when he said, "If I have been privileged to see farther than others, it's because I stood on the shoulders of giants."

It's only by asking questions, challenging the assumptions and the "truths" taken for granted at any given time, that science progresses. What if that turned out to be true about our personal lives, our individual growth and progress?

Guess what? It is true. When you break free of your assumptions about yourself, you will grow more than you ever thought possible.

You can never come to a conclusion about life. Life is an eternal thing just as we are an eternal thing. We have to start searching for more meaning of what we are. Well, the meaning of what we are has yet to be discovered by us.

—Ramtha

Bring It on Home

Pondering the Great Questions is a wonderful way to spend "quality time" with your mind. When was the last time you took your mind on a wild ride of mystery? Tried to get to the other side of Infinity?

Asking questions also has enormous practical value. It's the gateway to change.

For instance: Ever ask yourself, as Joe Dispenza asks, "Why do we keep recreating the same reality? Why do we keep having the same relationships? Why do we keep getting the same jobs over and over again? In this infinite sea of potentials that exist around us, how come we keep recreating the same realities?"

Or as Einstein put it, one of the definitions of insanity is to do the same things over and over and expect a different result.

That's where asking Great Questions comes in. They are *Great* because they open us up to a greater reality, a greater vista and greater options. And they come in the form of *Questions* because they come from the other side of the Known. And to get there is to change.

Ponder These for a While . . .

A note about "Ponder These": Some of these questions many of us can answer easily. But the trick is to not just look at the obvious, but to look at the unobvious—the subconsciousness. The place we don't look very often, if ever. When you consider the questions here, remember to look all the way inside yourself. Think about things that you may have picked up when you were young. Like fear, for instance: Does a fear of dogs permeate throughout your consciousness in other ways? Take some time. There's no one at the back of the room with a stopwatch!

- Remember your first three questions from the beginning of the chapter? What are they now?

- A spaceship lands next to you, and inside is *The Universal Book of Everything.* You get a bonus question, a just-for-fun question. What is it?

- And the bonus bonus: Are we back where we started? Or have we moved on?

Remind yourself of these questions as you read this book. They are bound to evolve as you evolve. That's the fun part! Keep a journal so you can watch your own evolution and remember.

All great things are achieved in a light heart!
—Ramtha

What

Who am I

Why an I... this time

WHO'S ASKING

...have got to do with it?

What are the...

...supposed I kn...

Do I have a soul, an inner spirit
Will it go on after my body dies?
What is con... to go
Where will it go?

WHY IS THERE
ANYTHING AT ALL

SCIENCE AND RELIGION: THE GREAT DIVORCE

In formal logic, a contradiction is the signal of defeat:
but in the evolution of real knowledge, it marks the
first step in progress toward victory.

ALFRED NORTH WHITEHEAD

Spirit and science are humanity's two grand approaches to The Truth. Both are searching for the truth about us and our universe; both seeking answers to the Great Questions. They are two sides of the same coin.

United in Their Source

O ur earliest known civilization, ancient Sumer (3800 B.C.E.), saw the pursuit of understanding the world around us and the world of the spiritual as the same thing. There was a god of astrology, a god of horticulture and a god of irrigation. The temple priests were the scribes and technologists investigating these fields of knowledge.

The Sumerians knew about the 26,000-year cycle, the precession of the equinoxes, the mutating of plants to produce fruits and vegetables, and an irrigation system that fed the entire "fertile crescent" (Tigris/Euphrates River basin).

Forward 3,000 years to ancient Greece. Philosophers were asking Great Questions like, Why are we here? What should we do with our lives? They also developed the theory of the atom, studied celestial movements and sought universal principles of ethical behavior. For thousands of years, the only study of the heavens was astrology. From astrology came modern-day astronomy. From astronomy came mathematics and physics. Alchemy, the search for transmutation and immortality, spawned the science of chemistry, which later specialized into particle physics and molecular biology. Today the search for immortality is carried on by the DNA biochemists.

In the mystic sense of the creation around us, in the expression of art, in a yearning toward God, the soul grows upward and finds the fulfillment of something implanted in its nature. . . . The pursuit of science [also] springs from a striving which the mind is impelled to follow, a questioning that will not be suppressed. Whether in the intellectual pursuits of science or in the mystical pursuits of the spirit, the light beckons ahead, and the purpose surging in our nature responds.

—Sir Arthur Eddington, astrophysicist, in *The Nature of the Physical World*

A World Alive

The world that people believed in before the Scientific Revolution was *alive*. In China, people saw the world as a dynamic interplay of energetic forces that are constantly in flux. Nothing is fixed and static; everything is flowing, changing or forever being born.

People in the West believed that the world at large expressed the will and intelligence of a Divine Creator. Its component parts were linked in a "Great Chain of Being," stretching from God through angels to man, animals, plants and minerals, all of which had their proper place in a living whole. Nothing stood alone; every part was related to every other part.

Native peoples on every continent lived in harmonious relationship with their surroundings—the animals and plants, the sun and rain, the living Earth. They often expressed this perception by finding "spirits" in mountains, streams and groves of trees, and based their religion and their science on learning to live in a way that pleased those spirits of the Earth and sky.

The goal of science in all these cultures was to gain knowledge in order to harmonize human life with the great forces of the natural world and the transcendent powers that all cultures sensed behind the physical world. People wanted to know how nature works, not in order to control and dominate it, but to live in accord with its ebb and flow. As the physicist and philosopher Fritjof Capra wrote in *The Turning Point,* "From the time of the ancients the goals of science had been wisdom, understanding the natural order and living in harmony with it. Science was pursued 'for the glory of God,' or, as the Chinese put it, to 'follow the natural order' and 'flow in the current of the Tao.'"

All this changed radically, starting in the middle of the 16th century.

In the greatest cultures of the ancient world there was a stairway between the human and the divine. The Earth and the cosmos were addressed as "thou," not "it." People felt they participated in a great cosmic mystery of which they were a part. People experienced the divine as imminent in the material world. Nature and the cosmos were ensouled with divine presence. Ceremonies like those performed at Stonehenge . . . connected Earth with heaven and strengthened the sense of participation in a divine reality.

—Anne Baring

Challenging the Power of the Church

In medieval Europe, the Church held a position of supreme power. Kingmaker, landowner and purveyor of the truth, the Church took it upon itself to be the one knower of everything. Its dogma was law, and its power was absolute. Not only were they legislating the way the spiritual world worked, in terms of heaven, hell and purgatory, they were also telling the physical Universe how to behave.

In 1543, Nicolas Copernicus had the audacity to challenge the Church and the Bible. He published a book suggesting that the sun, not the Earth, was the center of our universe. The Church did the most logical thing when confronted with the notion that it might be wrong—it forbade its followers from reading it. It placed his work on its "Index" of forbidden books and, remarkably, did not remove it until 1835!

Luckily for Copernicus, he died of natural causes before the Church could get to him. Two scientists who supported his work did not get off so easily. Giordano Bruno confirmed Copernicus' calculations, and speculated that our sun and its planets might be just one of many such systems in an endless universe. For this terrible blasphemy, Bruno was brought before the Inquisition (which is *still* a department in the Church), condemned as a heretic and burned to death.

Galileo Galilei also supported Copernicus' model. He, too, was called before the Inquisition, but because he was a personal friend of the pope, he was merely locked under house arrest (at the age of seventy) until his death. It's good to have friends in high places.

Galileo is often called the "father of modern science" because he was the first to base his work on the two pillars that have characterized the scientific enterprise ever since: empirical observation and the use of mathematics.

Because of Galileo's discoveries in the early 1600s, knowledge was no longer the property of the priesthood. Its validity would not be based on ancient authorities or ecclesiastical

As a young boy I pondered God a lot. I was told that God was beyond me and was a mystery that could never be fathomed. Being both arrogant and inquisitive, I decided that they were wrong. There had to be a way, I thought. When I discovered science in my teens, I became so excited. Even though I knew science was studying the aftereffects of a higher order, I felt that the things I learned came closer to the wonder of life than many of the dry moments I experienced as a child in church. When I found out about quantum mechanics, I was in heaven! (Pardon the pun.) Here was a language that I thought might start to explain the divine, and the idea of the observer might suggest that the divine is us. Science and spirit are not so different. They are different disciplines trying to understand the same thing.

—MARK

hierarchies. Rather, knowledge was to be gained through open inquiry and observation, and validated by agreed-upon principles, which soon became known as the *scientific method*.

The scientists did not pick battles with the Church. They knew it was hopeless and dangerous. Rather than attempt to formulate mathematical laws about God, the soul or even human nature and society, they restricted their activities to probing the mysteries of matter.

For its part, the Church did everything within its power to shut them down, to prevent the spread of ideas that might threaten its authority. But what the Church dreaded is precisely what happened. As the scientists persevered in the adventure of discovery, sent back dispatches from the frontiers of the known and used the growing body of knowledge to create ever-more-powerful technologies, the charm of the scientific enterprise drew increasing support.

Descartes Divides Mind from Body, Humanity from Nature

The 17th-century French philosopher and mathematician, René Descartes, widened the gap between science and spirit. "There is nothing included in the concept of body that belongs to the mind," Descartes said, "and nothing in that of mind that belongs to the body."

And thus the axe fell. The same coin (reality) was split down the middle. If spirit and science were having a divorce, Descartes was the lawyer who made it palatable.

Although Descartes believed that both mind and matter were created by God, he viewed them as completely different and separate. The human mind was a center of intelligence and reason, designed to analyze and understand. The proper domain of science was the material universe—nature—which he saw as a machine that operated according to laws that could be formulated mathematically. To Descartes, a great lover of

The riff between science and spirit affects us today because the scientists who are involved in this sort of debate know very little about the true teachings of spirit. They simply take the characters that they find retailed from every pulpit throughout the land and take this as a scientific spirit when in fact it's only a version of the science of the spirit. And, unfortunately, the churchmen don't know their science either, so the two sides are actually firing at cross-targets. These are simply two complimentary ways of looking at reality.

—Miceal Ledwith

clocks and mechanical toys, all things in nature, not only inanimate objects like planets and mountains, shared this mechanical nature. All the operations of the body, too, could be explained in terms of the mechanical model. He wrote, "I consider the human body as a machine." The separation of mind from body that Descartes made into a fundamental rule of science has caused endless problems, as we will see.

Francis Bacon and the Domination of Nature

Francis Bacon, a British philosopher/scientist, was also very instrumental in establishing the scientific method, which we can diagram like this:

Hypothesis → research and experimentation → draw general conclusions → test those conclusions by further research

Of course, this method has resulted in tremendous advances for humanity, from the pure delight of greater understanding of nature to improvements in health, engineering, agriculture, etc., to the first baby steps of space exploration. But that's only half of the story.

As Fritjof Capra has pointed out, Bacon viewed the scientific enterprise in terms that were "often outright vicious." Nature had to be "hounded in her wanderings," "bound into service," and "made a slave." The job of the scientist was to "torture nature's secrets from her." Unfortunately, this attitude that sought to extract knowledge in order to control and dominate nature (described as a "her") has become a guiding principle of Western science. Bacon summed it up in a phrase we all learned in school: "Knowledge is Power."

Newton's Classical Model

The person we most often closely associate with the formulation of the scientific worldview is Sir Isaac Newton, and the

> I spent most of my life with my head in the sand. Waking up preoccupied with what shoes I was going to wear was my safety net. I could never reconcile the notion of a guy up in heaven judging me, and I never could buy wholeheartedly the notion that I came from an ape. It always seemed to me there must be something else. But it was too big for little me to consider. So for a long time I left it to "the smarter people." Now I realize that unless I wake up and become a participant in this dialogue, science and religion will continue down their roads of elitism, dogma and power mongering. I think they need a good relationship therapist—US!
>
> —BETSY

mechanistic model of the world is often referred to as "Newtonian physics" or "the Newtonian model." These terms are justifiable, as Newton took giant steps beyond his predecessors, synthesized their ideas and methods, and advanced them greatly. The conclusions he came to, and the mathematical proofs he provided, were so powerful that for nearly 300 years scientists the world over were convinced that they described precisely how nature works.

Newton, like Descartes, saw the world as a machine, operating in three-dimensional space, with events (like the motions of the stars or the falling of apples) taking place in time. Matter was solid, with tiny particles at its core; these particles, as well as giant objects like planets, moved according to laws of nature, such as the force of gravity, which could be described with such mathematical precision that if we knew the initial conditions of any object, such as the whereabouts of a planet and the speed and pattern of its orbit, we could predict its future with absolute certainty. Newton's linking together of two such disparate events, the falling of an apple and the motion of a planet, was utterly revolutionary. The linking was mediated by a "force," in this case, gravity.

The mechanistic approach was soon applied to all the sciences: astronomy, chemistry, biology, and so on. With few variations (such as a more sophisticated view of the atomic level of reality), it's the world we all were brought up to believe in.

Newton and Religion

Consider this: As revolutionary as Newton and his colleagues were in their work, when it came to religion they did not question the dominant worldview of their age. They were immersed in it. Although they were responsible for initiating a radical new paradigm that would challenge and overturn understandings that had endured for centuries, they lived their personal lives very much in the midst of the medieval world into which they were born.

In the seventeenth century we came out of a period of time where we saw the universe as a living, vibrant entity, to the view of the world as a machine. Descartes and Newton solidified that concept, by using science and mathematics to describe a nonliving world of inanimate objects. They made some very beautiful calculations and enhanced our understanding of nonliving systems. Descartes looked at the world as a machine. He was very interested in clocks. The problem is, he and the other early scientists applied the model of a clock or a windup toy soldier to living systems. The idea was that if we understood the parts, the different components of the system well enough, we would understand how the whole system works. That may be true of a clock, but the problem is, we're not a machine at all, or a clock or a toy soldier.

—Daniel Monti, M.D.

Like other people, they believed that God was the master architect and builder of the world. Newton wrote in his major scientific work, *Principia Mathematica,*

> This most beautiful system of the sun, planets, and comets could only proceed from the counsel and dominion of an intelligent and powerful being. . . . This Being governs all things, not as the soul of the world, but as Lord over all. . . . He is eternal and infinite, omnipotent and omniscient. . . . He governs all things and knows all things that are or can be. . . .
>
> Why there is one body in our system qualified to give light and heat to all the rest, I know no reason but because the Author of the system thought it convenient.

As if to prepare the ages to come against the materialistic philosophy that would dominate Western thought in the name of Newtonian mechanics, Sir Isaac wrote, "Atheism is so senseless and odious to mankind that it never had many professors."

A Bitter Divorce

It was only later generations of scientists, focused entirely on the world machine, who found they had no need for God or spirituality. Set free of the constraints of religious dogma, the scientists reacted with a vengeance, proclaiming everything unseen and nonmeasurable to be fantasy and delusion. Many became as dogmatic as the Church authorities, declaring with self-righteous certainty that we are strictly little machines running around in a predictable machine universe governed by immutable laws.

The followers of Darwin provided the final stroke in the materialist triumph. Not only is there no God, and thus no creative intelligence guiding the unfolding of intergalactic life, but we ourselves, once at the center of the world, are nothing but random mutations, carriers of DNA's relentless quest for more, in a meaningless universe.

Hope for Reconciliation?

The separation of mind and body that Descartes made into a fundamental rule of science, and which scientific discovery believed for hundreds of years, has caused endless problems. By viewing the world outside our minds as nothing but lifeless matter, operating according to predictable, mechanical laws and devoid of any spiritual or animate quality, it divided us from the living nature that sustains us. And it provided humanity with a perfect excuse to exploit all "natural resources" for our own selfish and immediate purposes, with no concern for other living beings or for the future of the planet.

And the planet suffered. Raped of resources and stripped of purity, our polluted home began to spin toward the brink of extinction.

As science dug even deeper into its dead universe, it stumbled upon, and unlocked, a mystery. In the early years of the 20th century, the stranglehold of materialism was being cracked open by scientists like Albert Einstein, Niels Bohr, Werner Heisenberg, Erwin Schrödinger and other founders of quantum theory, who told the world: Probe deeply enough into matter, and it disappears and dissolves into unfathomable energy. If we follow Galileo and describe it mathematically, it turns out it is not a material universe at all! The physical universe is essentially non-physical, and may arise from a field that is even more subtle than energy itself, a field that looks more like information, intelligence or consciousness than like matter.

Two Sides of the Same Coin

At this late date, the coin remains split, with religion on one side and science on the other. Why? Not because reality is split,

If science and spirit are investigating the nature of unlimited reality—and, obviously, the more unlimited it is, the closer to reality—then they ought to eventually cross paths. The oldest known scriptures, the Vedas, talk about the physical world as illusion, maya. Quantum physics says reality is not the way we see it; rather, it is at best mostly empty, but really more like waves of insubstantial no-thing.

The Tibetan Buddhists talk about everything as "interdependent origination." In physics there is entanglement, which says all particles are connected, and have been since the big bang (where they got entangled in the first place). And more poetically we have in Zen their famous koan: "What is the sound of one hand clapping?" which is echoed by the physics question: "How can a particle be in two places at once?"

Professionals on both sides of the fence have dug into their respective disciplines, yet the history of human progress shows that evolution comes about by including wider and wider areas of study and integrating them.

What is the sound of two adversaries kissing?

—WILL

but because the adherents of their worldview are *people.* Remember why people don't ask Great Questions? Because the answer they get may not be what they want it to be.

What if the mind and matter are not split? What if there are observable feedback loops between the two? It's the 21st century, yet mainstream science still refuses to look at this.

Dr. Dean Radin, head scientist at the Institute of Noetic Sciences, has been pursuing the investigation of psychic phenomena with strict adherence to the scientific method. Even so, he still meets resistance within the mainstream scientific community.

As Dr. Radin says,

> They [mainstream scientists] have personal, private beliefs that have developed because of their experience, but they don't talk about it in public because in public, at least within the academic world, you're not supposed to talk about it. And this is one of the few areas in academia where this taboo is not only strong, but it has persisted for at least a century. I know many, many academic colleagues . . . distinguished people in their fields—in psychology, cognitive neuroscience, basic neurosciences, physics . . . who privately are very, very interested in . . . psychic phenomenon. Some of them are getting successful results in their experiments. Well, why aren't we hearing about it? Because the culture in the academic world says you cannot talk about it. So we're living in the parable of the emperor's new clothes. I mean, even at this point the taboo is so strong you're not even supposed to talk about the taboo. It's like a highly secret government project where the fact of the existence of the project is secret. Well, the taboo is secret; no one's supposed to talk about it. Once the taboo is addressed, that's the first stage in making it dissolve, and at that point, you will find an enormous amount of interest in studying these things within mainstream science.

What Questions do YOU Want Answered?

Well?

Does prayer promote healing? Are you able to affect physical reality with your mind? Can you perceive things outside of space/time? Can a being walk on water? Does the Higgs Particle exist?

What!?

Theoretical particle physics predicts the existence of the Higgs Particle—the particle that gives mass to other particles. Hundreds of millions of dollars are being spent to build more powerful accelerators to find it. And yet we think that most of the citizens of planet Earth would rather know the first four questions.

Certainly answering those first four questions would have a massive (no pun) impact on how we see ourselves and the world. Much more so than finding yet another particle. But the established world of science does not want to look at something that is thought to be "outside of their domain." Funny, because that's where the breakthroughs come from.

So, who now hijacked the search for truth?

Two sides of the same coin.

First the Church, and now the new priesthood—the Scientists.

THE HIGGS PARTICLE

The Higgs Particle is a theoretically predicted particle that gives all other particles in the universe mass. For decades scientists have been building bigger particle accelerators to find it because it is a heavy (massive) particle. They can't find the Higgs because it has too much mass, and it gives particles mass. So what gives the Higgs Particle mass then? Does this seem a little odd to anyone else out there? Maybe they should be looking for an "information" particle, one that in-forms particles as to their state (mass, charge, spin . . .)

Ponder These for a While . . .

- Have you hijacked your own search for the truth?
- What does spirituality mean to you?
- What is the difference, if any, between a dogma and a natural law?
- What are the dogmas in your own life?
- How do they govern how you perceive yourself and your reality?
- Do you use the scientific method in your own life?
- How has the split of science and religion affected your life?
- What is the difference between science and religion?
- How has dualism affected the way you perceive yourself and reality?
- Do you live your life as separate from nature and everyone else, or do you feel truly connected?
- How often do you feel like a lizard? Can you grow a tail?

PARADIGM SHIFT

*I am looking for a lot of men who have
infinite capacity to not know what can't be done.*

HENRY FORD

A paradigm is like a theory, but a little different. A theory is an idea that sets out to explain how something works, like Darwin's theory of evolution. It is meant to be tested, proved or disproved, supported or challenged by experiment and reflection. A paradigm, on the other hand, is a set of implicit assumptions that are not meant to be tested; in fact, they are essentially unconscious. They are part of our modus operandi as individuals, as scientists, or as a society.

Ever make Christmas cookies using a cookie cutter? No matter what the ingredients, they all come out of the oven looking pretty much the same.

—WILL

A paradigm is never called into question because nobody thinks about it. It's like having the proverbial rose-colored glasses on *all the time*; we see everything through those glasses. That's the reality we inhabit. All our perceptions come through that framework, and within that system are all the things we take for granted. We never question them—or even become aware of them—until we run into a wall and the rose-colored glasses are shattered, and suddenly the world looks different.

Paradigms and Belief Systems

Another way to understand a paradigm is as a belief system. If you have ever tried to define what your belief system is, what you value and believe, you know how hard it is. Maybe some of the issues you've thought about consciously aren't so difficult—you may believe in the importance of family, friendship, exercise, a healthy diet; you may have reason to believe your political affiliation is the sensible one, and so on. But there are dozens, maybe hundreds of unconscious, unexamined beliefs that run your life from the subterranean levels of shadowy awareness—beliefs about your worthiness and competence, for example, or whether people can be trusted or not—that were deposited in childhood and continue to determine how you relate to the world.

A paradigm is like the unconscious belief system of a culture. We live and breathe these beliefs, and we think and interact according to them.

The Old Scientific Paradigm Isn't Working

Practically every day, new scientific information is appearing that cannot be explained using the classical Newtonian model. Relativity theory, quantum mechanics, the influence of thoughts and emotions on our bodies, so-called "anomalies" like ESP, mental healing, remote viewing, people serving as mediums and channels, near-death and out-of-body experiences—all these point to the need for a different model, a new paradigm that would include all these phenomena in a more comprehensive theory of how the world works.

It's not just that the old model is insufficient to answer the questions the new research poses. An even more serious problem is that the old model has not done nearly enough to free human life of suffering, poverty, injustice and war. In fact, a good case could be made that many of these problems have grown *worse* because of the mechanical model that has long dominated our way of experiencing the world.

In beginning to understand which paradigms govern my life, I can start to see how I have created the situations in my life. Doing this movie and writing this book have broken up a huge unconscious paradigm of mine, the one that made me think, "I'm not that smart!" I never thought of myself as a person who could understand any of these concepts. Sure, I was savvy and crafty and could hustle my way around in the world and become successful. But I was not "book smart." My first week on this film, Will and Mark handed me about twenty books and said, "Start reading, because you're going to call these scientists and convince them to be in our film." It took a while for me to stop telling myself I couldn't do it. It was my job—I had to. And once I gave up my hold on my limitations, I dove in headfirst. Even now it still can haunt me—but then I just repeat, "I'm a genius!"

—BETSY

Repercussions of the Newtonian Paradigm

The materialist model of reality moved long ago from the ranks of "theory" to become set in stone as the implicit basis of all thought and research. It has governed scientific inquiry, and the scientific world's openness to what is possible or impossible, for 400 years. It tells us that the universe is a mechanical system composed of solid, material, elementary "building blocks." It asserts that what is real is what is *measurable*. And what is *measurable* is only that which we can perceive with our five senses, and any mechanical extension thereof. It also assumes that the only valid approach to gaining knowledge is to banish all feelings and subjectivity and become entirely rational and objective.

This way of relating to the world divides the wholeness of human life into mind and body. It declares feelings, passions, intuition and imagination as unworthy. It objectifies nature and sets us apart from it. In this view, nature becomes "resources" to control and exploit rather than an organic living system to care for and sustain.

According to the current scientific paradigm, we live in a mechanical universe that is a dead universe. It's the world of the machine. A living intelligence may have established it and set it in motion (as Newton and the early scientists firmly believed), but now, it is completely mechanical and predictable. Given any set of initial conditions, the outcome is completely determined. The effects are inevitable.

Now, even if the motion of planets is predictable like the falling of rocks and apples, and the behavior and relationship of objects in the material world are quantifiable (and we will see later that quantum physics has challenged these terms), to say this is true of human life is demeaning and stultifying. Where does this kind of life lead? If there's no freedom, if the path from wherever we are is completely determined—what's life about, then? There is no place for consciousness or spirit, for freedom and choice, in this model.

A New Paragm

In the words of Dr. Jeffrey Satinover, "A lot of people want quantum mechanics to be the rescuer from that kind of cold and pitiless indifference. And the reason people feel they need rescuing is because the cold, pitiless, mechanical idea is enormously powerful. Even if you don't profess to believe it, it has affected your life and the worldview of civilization to an immense degree."

Imagine yourself as a mechanical being (we've all seen enough science fiction movies to do this pretty easily) living in a totally dead world, in which all "things" are unconscious, unresponsive objects totally controlled by abstract laws of behavior. How does it feel? How do you feel about your loved ones, now that you are just a machine and that love is just a happenstance of brain chemistry, with nothing more than a evolutionary advantage to the DNA?

Do you believe that? And yet most of the scientists in the world are telling you that. They are the same people who tell you why the sky is blue, and why your car starts in the morning, and why trees convert carbon dioxide into oxygen. And if they had a big enough computer, they could tell you why you are sitting there right now reading this particular book. It's all initial conditions with which "you" (which is an illusion anyway) had nothing to do.

Do you believe *that*?

Of course, we have trouble thinking of *ourselves* as purely mechanical beings. That's because we're not. And neither is anyone else. We all experience that we have (or perhaps that we *are*) consciousness and spirit, and that we *do* make choices.

Or do we?

And here we are at the bottom of the paradigm rabbit hole. On the left is the life where we are conscious beings determining our path, and on the right is just ones and zeroes that somehow create the illusion of you.

From a classical point of view, we are machines, and in machines there is no room for conscious experience. It doesn't matter if the machine dies; you can kill the machine, throw it in the Dumpster . . . it doesn't matter. If that is the way the world is, then people will behave in that way. But there's another way of thinking about the world, which is . . . pointed to by quantum mechanics, which suggested that the world is not this clockwork thing but more like an organism. It's a highly interconnected organismic thing . . . which extends through space and time. So, from a very basic point of view having to do with morals and ethics, what I think affects the world. In a sense that's really the key for why the worldview change is important.

—Dean Radin, Ph.D.

Opposition from the Establishment

Just as in the time of Copernicus, Newton and the other 16th- and 17th-century pioneers of the scientific model, conservative elements in society are not only closed to this new knowledge, they are fiercely opposing it. The established orthodoxy is rigidly entrenched and unwilling to consider any change. These days, instead of the tried-and-true burning at the stake, the ecclesiastical authorities have been replaced by some (not all) people who use the power of universities, governmental grant-making institutions, and a closed-minded media to threaten the *livelihood* (through firing, denial of promotion or tenure, with-holding of grant money, ridicule and sarcasm) rather than the actual *lives* of "heretical" scientists whose ideas and research projects don't fit within accepted bounds.

Amit Goswami sees hope. He believes that opposition is not necessarily a bad thing. "An opposition has something significant to say. You never exclude things so long as you think that this is all rubbish, and just a cursory examination will eliminate the rubbish. But when things become significant and cursory examination doesn't do it anymore, that's when you become rigid and you want to exclude the other. Because the other is too dangerous. So the perception that the alternative scientists are making a good case of their data in their field is (impacting) already the world of establishment science. And that is why the polarization is a very good sign that we are getting somewhere."

The Evolution of Scientific Paradigms

One of the great truths about paradigms is that *they change*. Especially in science, which is an ongoing enterprise in which one generation builds on the work of those who came before, the paradigm of knowledge evolves as older views are proved to be incomplete or incorrect. Sometimes slow, sometimes kicking

For me the big paradigm change has to do with me (in here) and the rest of the universe (out there). If we are just little clocklike, windup soldiers in a ticktock universe, then why should it concern me what happens outside of me? And it's this attitude that makes it easy to bomb people, deplete resources and send future generations into a destitute world. Whereas when I extend my boundaries in time and space, it's all different. Practically, it means that instead of leaving the light on all night because I'm too lazy to turn it off, I think about the amount of coal or oil that gets burned lighting that bulb, then the manufacturing to make it, and the ozone hole and how in three generations there will be none left.

It's amazing how people will worry all day about their next meal, and yet not care about what someone will eat 100 years from now. The elk that graze in my meadow go from one patch to another, never eating ALL the seeds. They don't wear watches either.

—WILL

and screaming, but the grandeur of science is that it *does* move on! Science inexorably moves on, building a new view, a new structure on the foundation of the old.

Sometimes the current model bumps up against the ongoing march of knowledge and comes away bruised by the encounter. Then, whether with the support or against the opposition of the powers that be, the model gives way to a new one.

Dr. Hagelin has described the process this way:

> Within the progress of science, there are stages of understanding, stages of evolution of knowledge. Each of these stages brings its own worldview, its own paradigm, within which people act, within which governments are born, nations are born, constitutions are written, institutions are structured, education is created. So, worlds evolve from paradigm to paradigm as knowledge progresses. Each age has its own characteristic worldview, its characteristic paradigm, and one ultimately leads to another.

THE PARADIGM OF PARADIGMS

Along with many leading-edge scientists, William Tiller has run into in-the-box bias head-on. He says, "Okay, here we've done experiments with intention and they're very robust. . . . Why doesn't science rush in? That is the true sadness. The majority of scientists have become so locked into the conventional paradigm and the conventional way of viewing nature that they built a jail around themselves. If you provide experimental data that violates their precepts, they want it to go away, and so they'll sweep it under the rug. They will not let you publish. They will try to block all venues for communicating it because it's very uncomfortable. It is unfortunate—and it's always pretty much been that way. It's a human characteristic to be comfortable with a certain way of viewing the world. New stuff is uncomfortable;

I believe that the most far-reaching trend of our times is an emerging shift in our shared view of the universe— from thinking of it as dead to experiencing it as alive. In regarding the universe as alive and ourselves as continuously sustained within that aliveness, we see that we are intimately related to everything that exists. This insight . . . represents a new way of looking at and relating to the world and overcomes the profound separation that has marked our lives.

—Duane Elgin

you have to change your way of thinking."

Tiller explains one important reason the current paradigm of scientific reality needs to change:

"There is no place in . . . our present paradigm for any form of consciousness, intention, emotion, mind or spirit to enter. And because our work shows that consciousness can have a very robust effect on physical reality, that means that ultimately there must be a paradigm shift; a shift that would allow consciousness to be incorporated; the structure of the universe has to be expanded beyond what it presently is in order to allow it to enter."

Personal Paradigm Shift

The paradigm shift under way today is not just happening in science. It also extends into society and is powerfully impacting our culture. Perhaps the most important shift that is taking place is personal. Over the last couple of decades, many thousands, maybe many millions of people have undergone dramatic transformations in their values, perceptions and ways of relating to each other and to the world.

Why is this happening? One reason is that people have realized that at the end of their quest for flashier cars and bigger houses and shoes for every day of the year, what remains is an emptiness—the same emptiness they tried to fill with possessions and financial success. The materialistic worldview says: more money = better life. But having gotten more and finding that the emptiness remains, the conclusion is: The materialistic assumption is wrong.

Another reason? If the new paradigm is correct, and the universe is a living being, of which we, and our thoughts, and the planets, and all the subatomic particles are part, then the necessity for a new worldview will itself cause that to happen. It may be human arrogance (not that again!), which seems like *we* are bringing in the new view. A hungry organism always seeks out food. We are part of that organism, as are the planets, our thoughts and subatomic particles, and we are searching for a new way because we know we are camped out on death's doorstep.

And it's not a comfortable place to be. Polluted water and foul air. Overpopulation vying with starvation and suitcase-size weapons that can take down a city. The list goes on and on. Dr. Candace Pert says "the body always wants to heal itself." So if our reality, both physical and non-physical, is a huge organism, as is suggested by the "new physics," then that reality is this moment trying to heal itself. And out of that impulse new conceptions of the world are arising, even as old conceptions of the world fight to stay entrenched.

What is in the balance? Our notion of reality. Who is the balance? We are.

Ponder These for a While . . .

- What paradigm governs your reality?

- What color are your glasses (both conscious and unconscious)?

- How do you find the unconscious glasses?

- What is the predominant world paradigm?

- How is it different from your paradigm?

- How do they interact?

- Is *social consciousness* a paradigm?

- Is *People Magazine* a paradigm?

- Is the Bible?

- What would it take for you to shift to a new paradigm?

- Are you willing to give up everything attached to the old paradigm?

- What is your new paradigm?

- Is it *your* new paradigm or a new global paradigm?

- If we really are mutant machines . . . can you fall in love
  with your toaster?

WHAT IS REALITY?

What I thought was unreal now, for me, seems in some ways to be more real than what I think to be real, which seems now to be unreal.

FRED ALAN WOLF

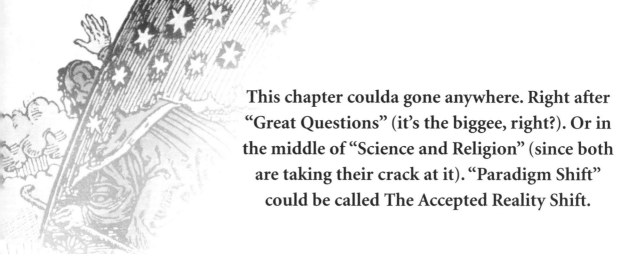

This chapter coulda gone anywhere. Right after "Great Questions" (it's the biggee, right?). Or in the middle of "Science and Religion" (since both are taking their crack at it). "Paradigm Shift" could be called The Accepted Reality Shift.

Animals and birds often live in a reality very different from ours. Some can hear sounds we can't hear, or see light frequencies (ultraviolet, infrared) that we can't see. Most mammals (like dogs) live in a world filled with scents, and rely much less than we do on vision. What about infants who stare for hours at an "empty" corner of the ceiling?

How about after "Sight and Perception" (the next chapter), which deals with what we perceive—and take to be real? Or the "Quantum Physics" chapter, which delves into reality at its core. Let's face it—it coulda/shoulda gone *everywhere*.

No? Tell me, *what is reality* when you've just fallen in love (the "Emotions" chapter) or when your true love has just died. What about the "Desire" chapter, which deals with choice and free will? Do you think those decisions are based on reality, or your assumption of it?

Let's see, what other upcoming chapters fit into our vision of reality? "Consciousness Creates Reality." Okay—there's a tie-in. On and on it goes . . .

This question *is* everywhere. It's in every chapter, in every moment that we live. Every decision is based on some construct of what is real to you. Yet when was the last time you took the rabbit-hole ride down into your assumptions about reality?

We asked more than a few scientists this question. In his response, Dr. David Albert touches on how and why we answer this question every day:

> If I get out of bed in the morning, okay, and I suddenly decide to take very seriously the claim, which is surely a

true claim . . . that I don't know for sure if my eyes are working correctly, so that for all I know even though it looks like there's a stable floor by the side of my bed, there might be a cliff or something like that. If I am unable to order those possibilities in terms of probabilities that I assign to them, then I'm not gonna get out of bed! Seems to me I'm paralyzed in the most literal sense of the word.

One hypothesis is there really is a floor there, and that's what I'm seeing. Another hypothesis is my seeing the floor is a hallucination, and there's a cliff there. By getting out of bed in the morning, you endorse one of those hypotheses as more likely than another. That's the way we're used to proceeding in our ordinary lives.

We endorse the reality that our eyes give us, so for us in that moment we answered the question looming above us—What is reality? Most people think reality is what our senses project to us. And, of course, science has gone along with that view for 400 years: If it is not perceivable by our five senses (or their extensions), it's not real.

But even this "reality" appears one way when we look at it with our eyes, and another if we look more deeply into it with a microscope or an atom smasher. Then it becomes totally different, unrecognizable.

And what about our thoughts? Are they part of "reality"? Take a look around right now. There are windows and chairs and lights and this book. You probably thought they were all real. All of them were preceded by an "idea" of windows and chairs. Someone imagined those windows and chairs and created them. So if the latter is real, is the idea real as well? Most people think thoughts and emotions are real—but when scientists explore "reality," they carefully avoid talking about such things.

What about consciousness, the fundamental fact of our own existence, that goes with us wherever we go? In order to do anything, to think, to dream, to create, to perceive, we have to be conscious. Isn't that part of reality? But where is it? What is it made of? Unlike material objects, intangible phenomena like consciousness can't be measured, but that doesn't mean they're not "real," or does it?

Many scientists are in a real pickle on this one. If it's real, then its reality can be examined; if it's not real, then they never have to go looking for it. And thus it will never be known as real.

So, "What is Real?"—possibly our most common question—isn't easy to answer. And yet who we are, what life is, what is possible and what is not, is all based on what we think is real.

—WILL

Back to the Laboratory!

I never questioned reality. Why would I do something as silly as that? Then the reality I was in became a mess, and I began to question my reality— not necessarily the tables and chairs, but my perception of it. Once I realized that my reality was only the construct of my limitations, I realized I had to be willing to dream outside of them. What is it that I truly desire that I don't believe I can have or become? The only thing "solid" in my reality is my perception of it. If I am willing to open my eyes to new possibilities, my reality can change.

—BETSY

Having not come up with the answer to "What is reality?"— which turned out to be way too big a question—humanity turned to the lab and tackled a simpler aspect: Take all the "stuff" around us, which we all pretty well agree is "real," and see what that's made of. That's much simpler than dreams or ideas or emotions or any of that inner stuff.

It was the Greek philosopher Democritus of Abdera who first had the idea of an atom: "Nothing exists except atoms and empty space; everything else is opinion." And that was a great place to start. So out came the electron microscopes and atom smashers and cloud chambers, and we big people peered into the world of the little things.

Now when you went to school, you probably were shown a model of an atom, with its solid nucleus and orbiting electrons, and you were probably told, "Atoms are the building blocks of nature." Nice try! It's a neat concept and diagrams quite nicely, but it just ain't so.

It turned out that those solid little atoms, in their neat little orbits, were really just energy packets. Then it was discovered that they're not really energy packets either, but momentary condensations of a field of energy. . . . Of course, as you know, every "atom" consists almost entirely of "empty space," so much so that it's a kind of miracle that we don't hit the floor every time we try to sit down on a chair. And since the floor is also mostly empty, where would we find something solid enough to hold us? The kicker here is that "we"—at least our bodies—are made up of atoms, too!

And now leading-edge research is suggesting that the so-called "empty space" within and between atoms is not empty at all; it's so lively with energy that one cubic centimeter—about a thimbleful or an area the size of a marble—contains more energy than all the solid matter in the entire known universe!

So what did you say Reality was?

Going Deeper

Long before the early Greek philosophers—and certainly long before quantum physicists—the sages of India knew that there was something important going on beyond the realm of the senses. Both Hindu and Buddhist seers taught, and still teach, that the world of appearances, the world we see with our senses, is *maya,* or illusion, and that something underlies this material realm, something that is more powerful and more fundamental, more "real" even though it's completely intangible. As so many spiritual texts suggest, there is a "higher reality" that is more fundamental than the material universe is, and it has something to do with consciousness.

This is precisely what quantum physics is revealing. It suggests that at the core of the physical world there is a completely non-physical realm, whether we call it information, probability waves or consciousness. And just as we commonly say that atoms are what things are "really" made of, if this view is correct, we would have to say that this underlying field of intelligence is, deep down, what the universe "really" is.

NASA astronaut Dr. Edgar Mitchell came to this conclusion on his return trip from space:

> In one moment I realized that this universe is intelligent. It is proceeding in a direction, and we have something to do with that direction. And that creative spirit, the creative intent that has been the history of this planet, comes from within us, and it is out there—it is all the same . . .
>
> Consciousness itself is what is fundamental, and energy-matter is the product of consciousness . . . If we change our heads about who we are—and can see ourselves as creative, eternal beings creating physical experience, joined at that level of existence we call consciousness—then we start to see and create this world that we live in quite differently.[1]

There is essentially nothing to matter whatsoever—it's completely insubstantial. The most solid thing you could say about all this insubstantial matter is that it's more like a thought; it's like a concentrated bit of information.

—Jeffrey Satinover, M.D.

---

[1] Which Dr. Mitchell did. He landed and created the Institute of Noetic Sciences (IONS) as a research institute to scientifically investigate his and others' "mystical" perceptions about reality.

The Possibly Greater/Truer Reality of Consciousness

I've noticed that some people sometimes think that asking questions like "What is reality?" is pointless and that it doesn't deal with day-to-day realities. But let's suppose for a moment that our world "out there" is constructed by our perception. How do we alter the foundation that is creating that world? Do we try and change the things "out there" in the world? Well, that's what most of us keep doing, and it never works. Ever tried to run away from a situation and find that your problems followed you? Well, of course they did. That's because you don't leave your nervous system behind. You still react the same way to the same old stimuli. So what's a better way? Know this: **Reality has everything to do with us.**

—MARK

Mitchell's realization parallels the experience of mystics throughout the ages, right up to today. Andrew Newberg, M.D., has studied the mystical/spiritual experience from the point of view of neuroscience and written about it in *Why God Won't Go Away: Brain Science and the Biology of Belief* and *The Mystical Mind: Probing the Biology of Belief.* He says that people who have a deep mystical experience and then "come back" to the ordinary world "still perceive that reality to be more real, to represent the truer, more fundamental form of reality; the material world that we live in is a more secondary reality for them."

Because of this, Dr. Newberg says, "We need to really look at the relationship between consciousness and material reality. . . . Whether or not the material world can actually be derived from a conscious reality or whether consciousness, itself, could even be the fundamental stuff of the universe."

Can We Ever Really Know?

In the 18th century, the German philosopher Emmanuel Kant pointed out that human beings can never truly know the nature of reality *as it is*. Our investigations only provide answers to the questions we ask, which are based on the capabilities and limitations of our minds. Everything we perceive in the natural world (whether with our senses or through science) comes through the filter of our consciousness, and is determined, at least to some extent, by the mind's own structures. Thus, what we see are "phenomena," that is, the interactions between the mind and whatever is "really out there." We don't see reality; we only see our construction of reality, built up by the neurons of our brains. The "thing-in-itself" is forever hidden from us.

To put it another way, science only gives us *models* of the world, not the world itself. As Miceal Ledwith says:

Well, you know, the quantum view of reality is not the be-all and the end-all. All we're trying to do in the history of science is to produce less and less imperfect models to express the nature of what exists and, of course, in maybe twenty or thirty years' time quantum physics will be replaced by a deeper and more profound understanding of reality, whatever that particular physics will be named.

And after science gives us those models, there is still the "us" to deal with, as Dr. Andrew Newberg points out:

As far as whether or not we're just living in a big holodeck, it's a question that we don't necessarily have a good answer to. I think this is a big philosophical problem that we have to deal with, in terms of what science can say about our world, because we are always the observer in science. We are always constrained by what is ultimately coming into our human brain that allows us to see and perceive the things that we do. So, it is conceivable that all of this is just a great illusion that we have no way of getting outside of to see what is really out there.

Levels of Reality

One piece of information that can be very helpful in dealing with mind-bending questions about the nature of reality is the idea that there are *different levels* existing simultaneously, and that all of them are real. In other words, the surface levels are real in their own right; it's only when we compare them to deeper levels that we say they are not really real; they are not the "ultimate" level. Arms and legs are real; cells and molecules are real; atoms and electrons are real. And consciousness is real. As Dr. John Hagelin says:

There literally are different worlds in which we live. There's surface truth, and there's deep truth. There's the macroscopic world that we see, there's the world of ourselves, there's the

I remember first running into the idea that we create reality in our minds, and that the "physical" world is just a construct. I had just read *Nature of Personal Reality* by Jane Roberts. So I closed my eyes and thought the wall in front of me wasn't real and that when I opened my eyes I would see through it. Didn't work. Or did it?

Sure, I had been holding the idea in my mind that the wall didn't exist, but everything about the way I sat there, expecting to be held up by the floor, being pulled by gravity, all supported the worldview that the world was the realest thing there is. When I get out of bed in the morning and put my feet down, I truly believe that the floor is real and not an illusion, and there's not really a bottomless pit there.

Every action we do assumes something about reality. Still, we seldom consciously ask it. We assume it, and reality complies, and so we never see the hand that makes it. There's a Zen question here somewhere. "What is the sound of one reality collapsing?"

—WILL

world of our atoms, the world of our nuclei. These are each totally different worlds.

They have their own language; they have their own mathematics. They're not just smaller; each is totally different, but they're complementary because I am my atoms, but I am also my cells. I'm also my macroscopic physiology. It's all true. They're just different levels of truth.

So:

1. It's all true.
2. None of it is true—there are only *models.*
3. *We* cannot ever get out of our own way to perceive the *all.*
4. By expanding our awareness, we can perceive the *all.*
5. *All* of the above are true.
6. *All* of the above are models.
7. *Or . . .*

Is Reality a Democratic Process?

In our day-to-day lives, in our moment-by-moment decision about reality, is it simply democratic? Or to put it another way, at what point in agreement from those around us does something become real? If there are ten people in a room, and eight see a chair and two see a Martian, who is delusional?

If twelve people see a lake as a body of water, and one person sees it as solid enough to walk on, who is delusional?

Going back a chapter, we could say a paradigm is simply the most commonly accepted notion (model) about what is real. We vote with our actions, and that becomes real.

So the kicker to all this is: Does consciousness create reality? Is that why no one has ever come up with a good answer— because reality IS the answer?

The easy answer to the question as to whether reality is illusory and it's really all fuzzy like all probabilities . . . would be yes. So if somebody came up to me and asked that question, I'd say yes, that's basically right. But it's more complicated than that because at the moment that you interact with reality, it does come into absolutely rock solid existence. It's only fuzzy when you're not interacting with it.

—Jeffrey Satinover, M.D.

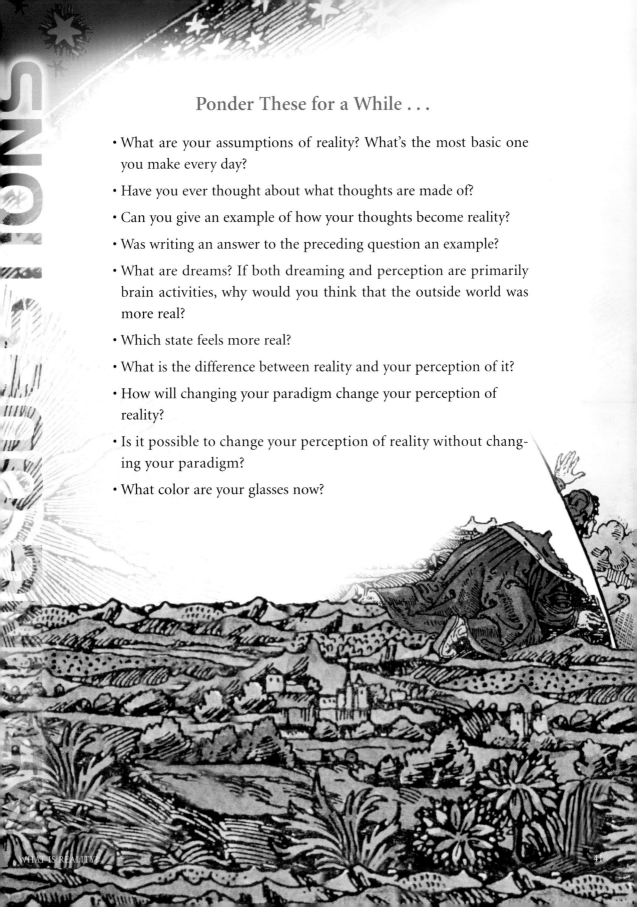

Ponder These for a While . . .

- What are your assumptions of reality? What's the most basic one you make every day?

- Have you ever thought about what thoughts are made of?

- Can you give an example of how your thoughts become reality?

- Was writing an answer to the preceding question an example?

- What are dreams? If both dreaming and perception are primarily brain activities, why would you think that the outside world was more real?

- Which state feels more real?

- What is the difference between reality and your perception of it?

- How will changing your paradigm change your perception of reality?

- Is it possible to change your perception of reality without changing your paradigm?

- What color are your glasses now?

This world
A dewdrop world
And yet . . .

—*Kobayashi Issa*

SIGHT AND PERCEPTION

The mind provides the framework, specific knowledge and specific assumptions for the eye to see. The mind constitutes the universe that the eye then sees. In other words, our mind is built into our eyes.

HENRYK SKOLIMOWSKI

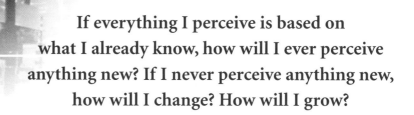

If everything I perceive is based on what I already know, how will I ever perceive anything new? If I never perceive anything new, how will I change? How will I grow?

Well after that trip down through that most nebulous and unknown of topics—Reality—it's good to get back to some real science. Tested, verified, accepted. And contradictory to the way most people believe, they perceive.

Who Sees What!?

Five nested levels of brain processing. That's what you just did to "see" each of these letters. It's not that your eyes just sent a picture to "you" of each letter. Your brain processed the visual data sent to it from your eyes to *construct* these letters.

The way it does this is to first break the incoming impulses into basic shapes, colors and patterns. Then it begins pattern matching with stored memories of similiar things, associating that with emotions and assigned meanings to events, tying this all together in an integrated "picture" and flashing that to the frontal lobe forty times a second. That's right. We don't even see continuously. It's like a flashing movie picture.

What this means is your brain paints everything you see. Let's say you're looking at a forest. Your brain is actually painting every leaf on every tree that you see. It paints that by linking to memories, or neuronets, of leaf, color, size, shape, and some-how placing them all together.

Top performing swimmers and divers, high jumpers, sprinters, weight lifters and other athletes have trained themselves to visualize their events in detail, using all the senses in order to simulate the total action they want to develop. This seemed very far out at first, especially to high-testosterone competitive athletes who couldn't quite fathom the training value of sitting still and closing their eyes, but by now it is absolutely proven to work and is commonplace.

This seems so outrageous and contrary to how we bounce around in the world. So how did neurophysiologists come up with this scheme?

Is That a Nose on Your Face?

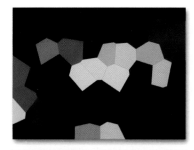

Scientists have learned how the brain builds visual images by studying stroke patients. Someone would have a stroke, and a small part of the brain would cease to work. Scientists would then see how that affected their visual sense.

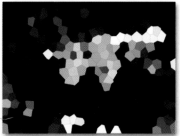

One example is of a person who had a stroke that wiped out a small part of the visual processing part of the brain. It was the part that (apparently) processed noses. This person could not see noses. They could see everything about a person, but if you walked up with a huge red Bozo the Clown nose and you asked them what was different about this person, they would never mention the nose. Even when prodded—"Boy, Bob sure has a nose!"—they would respond as if nothing were out of the ordinary. Everything else they perceived perfectly, so obviously the eyes were sending all the signals back. But for the nose, there was no one home. Instead of actually seeing the nose (Bozo or otherwise), they just saw what they thought the person's nose "should" look like.

Evidence that the brain actually perceives, not the eyes, is also shown on a less dramatic level: The place where the optic nerve goes through the eyeball to the back of the brain has no visual receptors. Therefore, you would expect that if we close one eye, we would see a black spot in the center. But we never do. That is because the brain paints the picture, not the eye.

More Data . . .

Scientists have found that if they measure the electrical output of a person's brain (using CAT or PET scans, for example) while they are looking at an object, and then again while they

We're bombarded by huge amounts of information, and it's coming into our body and we're processing it. It's coming in through our sense organs, and percolating up and up, and at each step we're eliminating information and, finally, what is bubbling up to consciousness is the one that's the most self-serving.

—Candace Pert, Ph.D.

What is it that I can't see that I want to see? How are my emotions affecting *and* effecting my perception of my reality? How can I perceive something new while I'm stuck in my old paradigm? What am I willing to change in order to perceive reality differently? How can changing my perceptions change my reality? Will it be better? Different? Both?

—BETSY

are *imagining* the same object, in both cases the same areas of the brain are activated. Closing the eyes and *visualizing* the object produce the same brain patterns as actually *looking* at the object.

Not only does the brain not distinguish between what it sees in its environment and what it imagines, it also does not seem to know the difference between an action performed and the same action visualized. This was first discovered by Edmund Jacobson, M.D. (creator of the Progressive Relaxation Technique for stress reduction) in the 1930s. When Dr. Jacobson asked subjects to visualize physical actions, he found very subtle muscle movements that corresponded to the movements the muscles would make in actually performing the activity. This information has been put to good use by competitive athletes around the world.

The Truth About Perception

Perception is a complex and many-faceted process that begins when our sensory neurons pick up information from the environment and send it, in the form of electrical impulses, to the brain. Like all living creatures, our sensory inputs are limited. We can't see infrared light, or sense electromagnetic fields as birds can (they use them for navigation). Nevertheless, the amount of information that comes pouring in from the five senses is staggering—somewhere on the order of 400 billion bits *per second.*

Obviously, we don't consciously receive and process anything near that amount—researchers say only about two thousand get through to our consciousness. So, as the brain gets to work "trying to create for us a story of the world," as Dr. Andrew Newberg puts it, "it has to get rid of a lot of extra data."

For example, as you read these words, even though your senses are picking up the temperature in the room, the feeling of your body on the chair, the texture of your clothing on your skin, the hum of the refrigerator and the smell of your shampoo, you're pretty much unaware of all that as you focus on the words of the book. Dr. Newberg continues:

> The brain has to screen out a tremendous amount of information that is really extraneous for us. It does that by inhibiting things. It does that by preventing certain responses and certain pieces of neural information from getting ultimately up into our consciousness, and by doing all of that, we ignore the chair that we're sitting in. That is, screening out the known. Then there's screening out the unknown . . .

> If we see something the brain can't quite identify, we grab onto something similar. ("It's not a squirrel . . . but it's something just like that.") If there's nothing close, or it's something we know to not be real, we discard it with, "I must have been imagining things."

So, we don't actually perceive *reality*; we see the image of reality that our brain has built up out of sensory input, plus countless associations drawn from the vast neural networks of our brain. "It depends on what your experiences have been," Dr. Newberg says, "and how you ultimately process the information, that really creates your visual world . . . The brain is what ultimately perceives reality and creates for us our rendition of the world."

Your brain doesn't know the difference between what's taking place out there and what's taking place in here.

—Joe Dispenza

There is no "out there" out there, independent of what's going on "in here."

—Fred Alan Wolf, Ph.D.

Emotions and Perceptions

Dr. Pert's research at the National Institutes of Health suggests that it is not only what we believe is real, but also how we feel about what our senses are picking up, that determines how and if we are going to perceive it. She says, "Our emotions decide what is worth paying attention to . . . The decision about what becomes a thought rising to consciousness and what remains an undigested thought pattern buried at a deeper level in the body is mediated by the receptors."

Our eyes are moving all the time. They are moving over this whole field of energy—so why do they focus and start to let in one area and not let in another? It's pretty simple: We see what we want to believe. And we turn away from things that are too unfamiliar or unpleasant.

—Candace Pert, Ph.D.

As Joe Dispenza puts it: "Emotions are designed to reinforce chemically something into long-term memory. That's why we have them." Our emotions are linked in at a low level of visual processing, somewhere close to the first step. This makes sense from an evolutionary point of view. If you are walking down the path and a tiger jumps in front of you, you will process that picture and start running before you realize why.

Four hundred billion bits a second. Even once we throw out the stuff that is "unreal" (Martians) and the stuff that is "irrelevant" (smell of shampoo), there is still a lot of bits left. Emotions give them their relative weighting or importance. They are a hardwired shortcut to perception. They also provide us with the unique capability to not see what we simply *don't want to see.*

Paradigm and Perception

So, if we build reality out of our already-existing store of memories, emotions and associations, how do we ever perceive anything new?

The key is new knowledge. By expanding our paradigm, our model of what is real and what is possible, we add new options to the list our brains carry around. Remember, that list is just a working description of reality based on our own personal experience; it's not reality itself. New knowledge can open our minds to new types and levels of perception and experience.

New information is important, but complete knowledge involves both understanding and *experience.* If you want someone to know what it's like to eat a peach, you can give them information about it—"It's juicy, and sweet, and smooth . . ."—but they'll never really know until they bite into one. So to expand our paradigm and open ourselves to a bigger life, we also need new experiences.

For example, when was the last time you blew your own mind? The last time you did something so outrageously "not you" that you stood with your mouth open saying, "I can't believe I did that."

In *Journey to Ixtlan,* Carlos Castaneda recounts one of Don Juan's lessons: to "Stalk Oneself." In other words, to learn one's own habits like you were studying prey, so that you can trap yourself doing your habitual and do something totally new.

It's back to the old questions: If you perceive only what you know, how do you ever perceive anything new? If you create you, how do you ever create any new you?

Once it is realized that we are only able to experience life within the boundaries of what we already know, it becomes obvious that if we want to have a broader and richer life, if we want more opportunities for growth, achievement and happiness, we need to bust a move on ourselves by asking great questions, experiencing new emotions and packing more data into our neuronal nets.

We Create Our World

The bottom line, at least as far as science has gone up till now, is this: We create the world we perceive. When I open my eyes and look around, it is not "the world" that I see, but the world my human sensory equipment is able to see, the world my belief system allows me to see, and the world that my emotions care about seeing or not seeing.

Although we resist this notion and want to believe there's a "real world" that we can all perceive and agree on, the fact is that people often—and perhaps always—perceive the same things quite differently. For example, when a crime is recounted by several witnesses (as in the classic Japanese film, *Rashomon*), the versions of "what actually happened" differ widely—not only about the fine details of the crime, but even the appearance (such as hair color, height, dress) of both victim and perpetrator. Each witness to the crime believes he or she has the correct story—but what they really have is their own perception of what happened.

We are creating our world all the time in a myriad of ways.

I heard a story recently about an insurance company that was trying to analyze a particular problem in Saskatchewan. Turns out that light aircraft pilots who ran into engine trouble would try and land on the nearest freeway they could find that was relatively empty. One of the things that consistently happened is that once they landed and slowed down, they rarely left enough inertia to actually pull off the road. (They were probably so happy just to be alive.) Often drivers would smash into the aircraft and, when questioned by the police, they would almost always say that they never saw the plane. One moment they were driving; the next they hit something. The insurance company discovered that the reason this happens is because the last thing a driver expects to see on a freeway is a small aircraft, so they never see it.

—MARK

Sight and perception are the most obvious, scientifically veri-fiable way in which we do this. The question that is paramount is: Does it stop here? Is this the limit of our affecting the world we see?

One Step Beyond

Lest you think science has come to the end of raveling the mystery of sight, here's a plunge farther down the you-know-what hole.

Karl Pribram revolutionized the way people think of the brain by saying it was essentially holographic. He said that its processing was spread out all over the brain, and that like a holograph each part contained the whole. That was strange enough, but then he applied this model to how we perceive. He said the universe is essentially holographic, and the sole reason why we sense we are "in" reality, instead of just "perceiving" reality, is because our brain holographically links with the "out there" (in which case time and space go away),[1] and thus our perception is not just processed in the brain, but by moving outside of the brain to interact with "out there."

And that's why, no matter how good the virtual reality goggles are, they will never completely convince you that you are "in" that reality.

But if reality is holographic, is it possible to perceive that directly? Our senses are limited; they are cookie cutters press-ing into reality. Whereas explorers in consciousness report that it is possible to experience the world completely, directly, all the universe and a grain of sand, all at once. And from that point of view, everything—everything we perceive with the senses—is *maya,* illusion. So it's all just a point of view.

---

[1] Actually it's an alteration of frequency and phase relationships.

Ponder These for a While . . .

- How does your paradigm or attitude affect what you see?

- What emotional state do you find yourself in most often? How does that state affect your perceptions?

- Can you see anything that exists outside of that emotional state?

- If you perceive only what you know, how can you perceive anything new?

- What are you willing to do to perceive something new?

- Why don't you see auras?

- Where is that new perception coming from?

Ludwik Fleck, the Polish epistemologist and microbiologist who inspired Thomas Kuhn's notion of the paradigm, noticed that when beginning students are given microscopic sections to observe, at first they are unable to do so. They cannot see what is there.

On the other hand, they often see what is not there. How can this be? The answer is simple, because all perception, particularly sophisticated forms of perception, requires rigorous training and development. After a while, all students begin to see what is there to be seen.

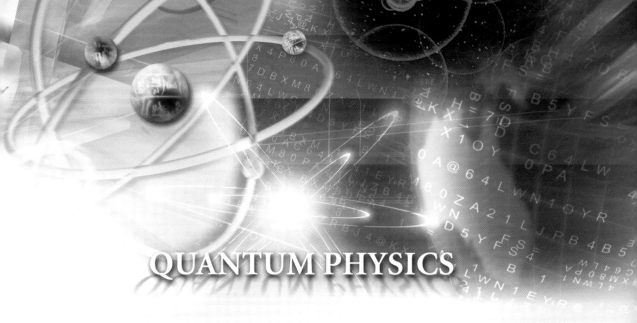

QUANTUM PHYSICS

I think I can safely say that nobody understands quantum mechanics.

RICHARD FEYNMAN

Awarded the 1965 Nobel Prize for the development of quantum electrodynamics

$$\frac{\eta^2}{2m}\frac{\partial^2 \psi(x,t)}{\partial x^2} + U(x)\psi(x,t) = i\eta\frac{\partial \psi(x,t)}{\partial t}$$

$$\frac{dA}{dt} = \frac{1}{i\hbar}[A, H]$$

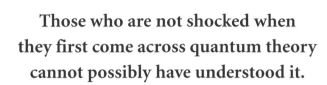

> Those who are not shocked when
> they first come across quantum theory
> cannot possibly have understood it.
>
> —Niels Bohr,
> **awarded the 1922 Nobel Prize for
> work on the structure of the atom**

I f folks like Nobel Prize winners don't understand quantum theory, what hope have we? What does one do when reality comes knocking at your door and tells you things that are confusing, baffling and puzzling? How you react, how you proceed in life and what your options are tells a lot about you, but that's a mystery for a later chapter. Right now, let's gossip about electrons, photons and quarks, and how something (if it is a *thing*!) so tiny could be so unfathomable and rip apart our well-ordered and understandable world.

On the one hand, this is an acutely paradoxical, puzzling, conceptually confusing theory. On the other hand, we have no option of throwing it out or neglecting it because it is the most powerful proven tool for predicting the behaviors of physical systems that we have ever had in our hands.

—David Albert, Ph.D.

The Known Meets the Unknown

Classical Newtonian physics was based on observations of the solid, everyday objects of ordinary experience, from falling apples to orbiting planets. Its laws were repeatedly tested, proven and extended over several hundred years. They were well understood and did a great job of predicting physical behavior—as seen in the triumph of the Industrial Revolution. But in the late 19th century, when physicists began developing the tools to investigate the very tiny realms of matter, they discovered something very puzzling: Newtonian physics did not work! It could neither explain nor predict the results researchers were finding.

Over the next hundred years, an entirely new scientific description has grown up to explain the world of the very small. Known as quantum mechanics or quantum physics (or simply quantum theory), this new knowledge does not *replace* Newtonian physics, which still works quite well to explain large, macroscopic objects. Rather, the new physics has been invented to (boldly) go where Newtonian physics could not go: to the subatomic world.

"The universe is very strange," says Dr. Stuart Hameroff. "There seem to be two sets of laws that govern the universe. In our everyday, classical world, meaning at roughly our size and time scales, things are described by Newton's laws of motion set down hundreds and hundreds of years ago. . . . However, when we get down to a small scale, to the level of atoms, a different set of laws takes over. These are the quantum laws."

The term quantum was first applied to science by the German physicist Max Planck in 1900; it is a Latin word that simply means amount or quantity, but is used to mean the smallest unit of any physical property, such as energy or matter.

Fact or Fiction?

What quantum theory has revealed is so mind-boggling that it sounds more like science fiction: Particles may be in two or more places at once. (A very recent experiment found that one particle could be in up to 3,000 places!) The same "object" may appear to be a particle, locatable in one place, or a wave, spread out over space and time.

Einstein said that nothing can travel faster than the speed of light, but quantum physics has demonstrated that subatomic particles seem to communicate *instantaneously* over any expanse of space.

Classical physics was *deterministic*: Given any set of conditions (such as the position and velocity of an object), you could determine with certainty where it would go. Quantum physics is *probabilistic*: You can *never* know with absolute certainty how a specific thing will turn out.

If you want to put your finger on one of the profound philosophical shifts between classical mechanics and quantum mechanics, [it is that] classical mechanics is built from the ground up around what we now know is a fantasy: the possibility of observing things passively . . . Quantum mechanics put a decisive end to that.

—David Albert, Ph.D.

Classical physics was *reductionist*: It was based on the premise that only by knowing the separate parts could you eventually understand the whole. The new physics is more organic and *holistic*; it is painting a picture of the universe as a unified whole, whose parts are interconnected and influence each other.

Perhaps most importantly, quantum physics has erased the sharp Cartesian distinction between subject and object, observer and observed, that has dominated science for 400 years.

In quantum physics, the observer *influences* the object observed. There are no isolated observers of a mechanical universe, but everything *participates* in the universe. (This is so important that we will talk about it in a separate chapter.)

Mind Boggle #1—Empty Space

Let's start with something familiar to most of us. One of the first cracks in the structure of Newtonian physics was the discovery that atoms, the supposedly solid building blocks of the physical universe, were mostly made up of empty space. How empty? If we use a basketball to represent the nucleus of a hydrogen atom, the electron circling it would be about twenty miles away—and everything in between would be empty. So as you look around, remember that what really is there are tiny, tiny points of matter surrounded by nothing.

Well, not really. That supposed "emptiness" is not empty at all: It contains enormous quantities of subtle, powerful energy. We know energy increases as we go to subtler levels of matter (nuclear energy being a million times more powerful than chemical energy, for example). Scientists now say there is more energy in one cubic centimeter of empty space (about the size of a marble) than in all the matter of the known universe. Although scientists have not been able to measure this directly, they have seen effects of this sea of immense energy.[1]

---

[1] For more information on this, look up "Van der Waals Forces" and the "Casimir Effect."

Mind Boggle #2—Particle, Wave or Wavicle?

Not only is there "space" between particles, but as scientists probed more deeply into the atom, they found that the subatomic particles (the constituents of the atoms) are not solid either. And they appear to have a dual nature. Depending on how we look at them, they can behave as either particles or as waves. Particles can be described as separate, solid objects with specific locations in space. Waves, on the other hand, are not localized or solid, but are spread out, like sound waves or the waves in water.

As waves, electrons or photons (particles of light) have no precise location, but exist as "probability fields." As particles, the probability field "collapses" into a solid object locatable in a specific place and time.

Amazingly, what seems to make the difference is observation or measurement. Unmeasured, unobserved electrons behave as waves. As soon as we subject them to observation in an experiment, they "collapse" into a particle and can be located.

How can something be both a solid particle and a soft, flowing wave? Perhaps the paradox can be resolved by recalling what we said above: Particles *behave* as a wave or particle. But the "wave" is just an analogy. Just like "particle" is an analogy from our everyday world. This wave notion was solidified into quantum theory by Erwin Schrödinger, who with his famous "wave equation," summed up mathematically the wave-like probabilities of the particle before observation.

In an attempt to make it clear they don't really know what the BLEEP they're dealing with, but that whatever it is they've never seen anything like it, some physicists have decided to call this phenomenon a "wavicle."

Mind Boggle #3—Quantum Jumps and Probability

In studying the atom, scientists found that when electrons move from orbit to orbit around the nucleus, they do not move

DOWN THE PARTICLE RABBIT HOLE

As Schrödinger was formulating his wave equation, Werner Heisenberg solved the same problem using advanced matrix mathematics. But the math was obscure, not relatable to our experience and didn't roll off the tongue like "wave," so "matrix transformations" were dropped in favor of the "wave" equation. It's all just an analogy.

For me this science is staggering in its implications. First of all, it behaves like magic. It behaves the why I used to think things behaved when I was a child. So what do we say of that young boy with his dreams and fantasies? Was he delusional? Perhaps. But suspiciously, quantum mechanics seems to be just as magical. The question is, where is the dividing line between the weird, wacky quantum world and the large-scale, seemingly solid world? Since I was a teenager, I always wondered to myself if I'm made up of subatomic particles that do really weird things . . . Is it possible I might be able to do really weird things?

—MARK

When a subatomic "object" is in its wave state, what it will become when it is observed and becomes located is uncertain. It exists in a state of multiple possibilities. This state is called superposition.

It is like flipping a coin in a dark room. Mathematically, even after it has landed on the table, we can't say whether it is heads or tails. As soon as the light goes on, we "collapse" the superposition, and the coin becomes heads or tails.

Like turning on a light, measurement of the wave collapses the quantum mechanical superposition, and the particle appears in a measurable, "classical" state.

through space the way ordinary objects move—rather, they move *instantaneously*. That is, they disappear from one place, one orbit, and appear in another. This was called a *quantum jump*.

As if that didn't break enough rules of commonsense reality, they also discovered that they could not determine exactly where the electrons would appear, or when they would jump. The best they could do was to formulate the probabilities (Schrödinger's wave equation) of an electron's new location. "Reality as we experience it is constantly being created freshly at every moment, out of this pool of possibilities," says Dr. Satinover, "but the real mysteriousness in this is that out of that pool of possibilities, which one is the one that is gonna happen is determined by *nothing that's part of the physical universe*. There is no process that makes that happen."

Or as is often stated: Quantum events are the only truly random events in the universe.

Mind Boggle #4—The Uncertainty Principle

In classical physics, all of an object's attributes, including its position and velocity, can be measured with precision limited only by our technology. But on the quantum level, whenever you measure one property, such as velocity, you cannot get a precise measurement of other properties, such as position. If you know where something is, you can't know how fast it's going. If you know how fast it's going, you don't know where it is. And no matter how subtle or advanced the technology, it is impossible to pierce this veil of precision.

The Uncertainty Principle (also referred to as Indeterminacy) was formulated by Werner Heisenberg, one of the early pioneers of quantum physics. It states that no matter how hard you try, you cannot get a precise measurement of both velocity and position. The more we focus on one, the more lost in uncertainty the measurement of the other becomes.

Mind Boggle #5—Non-Locality, EPR, Bell's Theorem and Quantum Entanglement

Albert Einstein did not like quantum physics (putting it mildly). Among other things he responded to the randomness described above with the infamous quote: "God does not play dice with the universe." To which Niels Bohr responded: "Stop telling God what to do!"

In an attempt to defeat quantum mechanics, in 1935 Einstein, Pedolsky and Rosen (EPR) wrote out a thought experiment that attempted to show how ludicrous it was. They cleverly drew out one of the implications of quantum that was not appreciated at the time: You arrange to have two particles created at the same time, which means they would be entangled, or in superposition. Then you shoot them off to opposite sides of the universe. You then do something to one particle to change its state; the other particle instantaneously changes to adopt a corresponding state. Instantaneously!

The idea of this was so ludicrous that Einstein referred to this as "spooky action at a distance." According to his theory of relativity, nothing can travel faster than the speed of light. And this was infinitely fast! Furthermore, the idea that an electron can keep track of another one on the other side of the universe simply violated every common sense of reality.

Then in 1964 John Bell created a theory that in effect said yes, the EPR assertion is correct. That is exactly what happens—the idea of something being local, or existing in one place, is incorrect. Everything is non-local. The particles are intimately linked on some level that is *beyond time and space.*

Over the years since Bell published his theorem, this idea has been verified time and again in the lab. Try and wrap your mind around this for a minute. Time and space, the most basic features of the world in which we live, are somehow superceded in the quantum world by the notion of everything touching all the time. No wonder Einstein thought this would be the death shot to quantum mechanics—it makes no sense.

The interesting question for me isn't, "Why is quantum so interesting?" but, "Why are SO MANY PEOPLE interested in quantum?" It defies our every notion about the way the world is; it tells us that the most obvious things that we KNOW are true, simply aren't. And yet it captivates millions of people who supposedly "don't have a scientific bone in their body."

—WILL

What is the sound of one electron collapsing?

Nevertheless, this phenomenon seems to be an operable law in the universe. In fact, Schrödinger was quoted as saying that entanglement was not *one* of the interesting aspects of quantum; it was *the* aspect. In 1975, the theoretical physicist Henry Stapp called Bell's Theorem "the most profound discovery of science." Note that he says *science,* not just *physics.*

Quantum Physics and Mysticism

It's probably getting easy to see why the two domains of physics and mysticism rub up against each other. Things being separated but always touching (non-local); electrons move from A to B, but never in between; matter appearing (mathematically) to be a distributed wave function and only collapsing, or becoming spatial existent, when measured.

Mystics seem to have no problems with these ideas, most of which predate the particle accelerator. Many of the founders of quantum had a major interest in spiritual matters. Niels Bohr used the ying/yang symbol in his coat of arms; David Bohm had long discussions with the Indian sage Krishnamurti; Erwin Schrödinger gave lectures on the Upanishads.

But does quantum physics *prove* the mystical worldview? If you ask physicists, you'll get answers all over the map. Ask this at a physics cocktail party and state one view emphatically, and you *may* (quantum is probabilistic after all) get a fistfight.

Aside from the dyed-in-the-wool materialists, the consensus seems to be that we are at the stage of analogy. That the parallels are just too amazing to ignore. That the mind-set to hold a paradoxical view of the world is the same in both quantum and Zen. As we quoted Dr. Radin before: "But there's another way of thinking about the world which is *suggested;* it's *pointed* to by quantum mechanics."

Questions about what causes the collapse of the wave function, or whether quantum events are truly random, are still largely unanswered. But while the urge to produce a truly unified concept of reality, which of necessity includes us, and which answers the quantum mysteries, is compelling, the contemporary philosopher Ken Wilber also urges caution:

> The work of these scientists—Bohm, Pribram, Wheeler and all—is too important to be weighed down with wild speculations on mysticism. And mysticism itself is too profound to be hitched to phases of scientific theorizing. Let them appreciate each other, and let their dialogue and mutual exchange of ideas never cease . . .
>
> My point, therefore, in criticizing certain aspects of the new paradigm is definitely *not* to forestall interest in further attempts. It is rather a call for precision and clarity in presenting issues that are, after all, extraordinarily complex.[2]

Conclusions

Conclusions? You've got to be kidding! Please, if you have some conclusions, run with them. Regardless, welcome to the contentious, exhilarating, puzzling, revelatory world of abstract thought. Science, mysticism, paradigms, reality—look what we humans have investigated, discovered and debated about.

Look at how the human mind has explored this strange world in which we seem to have found ourselves.

This is our true greatness.

We stand on the shoulders of billions of genetic lifetimes to give us this perfect genetic body, this perfect genetic brain that it took thousands of years to evolve, so that we could have this conversation in the abstract. If we are here to be embodied in the greatest evolutionary machine there ever was, our body and our human brain, then we have deserved the right for "what ifs."

—Ramtha

---

[2] Ken Wilber. *The Holographic Paradigm* (Boston: Shambhala, 1985).

Ponder These for a While . . .

- Think of an example of an experience of Newtonian physics in your life.

- Has Newtonian physics defined your paradigm?

- Does having new knowledge about the wacky, weird world of quantum change your paradigm? How?

- Are you willing to experience beyond the known?

- Think of an example of a quantum effect in your life.

- Who or what is the "observer" that determines the nature and location of the "particle"?

OBSERVER

Mirror mirror on the wall, who Collapses the very small?

QUANTUM VERSION OF AN OLD FAIRY TALE

> My conscious decision about how
> to observe an electron will determine the
> electron's properties to some extent. If I ask it
> a particle question, it will give me a particle
> answer. If I ask it a wave question,
> it will give me a wave answer.
>
> —Fritjof Capra

This profound shift in physicists' conception of the basic nature of their endeavor, and of the meanings of their formulas, was not a frivolous move: It was a last resort. The very idea that in order to comprehend atomic phenomena one must abandon physical ontology, and construe the mathematical formulas to be directly about the knowledge of human observers, rather than about the external real events themselves, is so seemingly preposterous that no group of eminent and renowned scientists would ever embrace it except as an extreme last measure.

—Henry Stapp

When faced with the experimental evidence that the process of observing appears to influence what is being observed, science was forced to drop four centuries of assumptions and grapple with a revolutionary idea—we *are* involved in reality. Although the nature and extent of this influence is still being hotly debated, it is clear that, as Fritjof Capra puts it, "The crucial feature of quantum theory is that the observer is not only necessary to observe the properties of an atomic phenomenon, but is necessary even to bring about these properties."

The Observer Affects the Observed

Before an observation or measurement is made, the object exists as a probability wave (technically called a *wave function*). It has no specific location or velocity. Its wave function or probability wave contains the likelihood that, when observed in a measurement, it will be *here* or *there*. It has potential positions, potential velocities—but we won't know what those are until it is observed.

"In this view," writes Brian Greene in *The Fabric of the Cosmos*, "when we measure the electron's position we are not measuring an objective, preexisting feature of reality. Rather, the act of measurement is deeply enmeshed in creating the very reality it is measuring." Fritjof Capra concludes: "The electron does not have objective properties independent of my mind."

All of this serves to blur the once-sharp distinction between the "world out there" and the subjective observer, which seem to merge or dance together in the process of discovering—or is it creating?—the world.

The Measurement Problem

Today this effect of observation is usually referred to as the measurement problem. While the early descriptions of this phenomena included the conscious observer, there have been many attempts to remove the troublesome word "conscious" from the problem. Questions about what is conscious quickly arose: If a dog looks at the results of an experiment on electrons, will that collapse the wave function?

In removing *consciousness* from the problem, physicists were able to realize the fact mentioned earlier: The fantasy of being able to make a measurement and not affect the measured was forever ruled out. The proverbial "fly on the wall" that sits there not influencing things cannot exist. (And we don't have to worry about whether or not the fly is *conscious*!)

To come to terms with the problems of observers, measurement, mind and collapse, numerous theories have been put forward over the years. The first, and still often discussed, theory is the Copenhagen Interpretation.

The Copenhagen Interpretation

The radical idea that the observer has an inescapable influence on any observed physical process, that we are not neutral,

I think one of the fantastic things people miss when talking about the observer is who that really is. Maybe we've become so used to the word that we haven't really understood it. The observer is every human being, regardless of sex, race, social standing or creed. That means that EVERY human being has the ability to observe and change subatomic reality. You could take anybody off the street—CEO, janitor, prostitute, violinist, policeman—and they could do it. Not just scientists in their hallowed halls. This science belongs to everyone because the science itself is a metaphor to explain US, the human being.

—MARK

> *To understand quantum mechanics completely, to determine fully what it says about reality . . . we must come to grips with the quantum measurement problem.*
>
> —Brian Greene
> *The Fabric of the Cosmos*

> *The question might be, can we make a mathematical model of what an observer is doing, when an observer is observing and changing reality? So far, that's eluded us. Any mathematical model that we use that brings in observation seems to introduce discontinuities in the mathematics. The observer has been left out of the equation of physics for a simple reason—it's easier to do things that way.*
>
> —Fred Alan Wolf, Ph.D.

objective witnesses to things and events, was first insisted upon by Niels Bohr and his colleagues in Copenhagen, where Bohr lived; thus it's often called the Copenhagen Interpretation. Bohr argued that Heisenberg's Uncertainty Principle implied more than just the fact that you cannot determine exactly both how fast a subatomic particle is moving and where it is located. Bohr's contention, as Fred Alan Wolf explains, was that, "It's not only that you cannot measure it. It isn't an 'it', until it's an *observed* 'it'. Heisenberg thought there were 'its' out there." He could not accept that there were no "its" until an observer was involved. Bohr believed that the particles themselves don't even come into existence until we observe them, and that reality on a quantum level does not exist until it is observed or measured.

Indeed, many scientists resisted and disputed this difficult and puzzling notion that goes against common sense and our ordinary daily experience. Einstein and Bohr argued into the night on numerous occasions, with Einstein saying that he simply couldn't accept it.

There *is* disagreement—some might say a raging debate—about whether or not this means that human consciousness, the human observer (as opposed to nonhuman), is what collapses the wave function and brings the object from a state of probability to its point value.

Heisenberg held that the mind was intrinsic to the problem. He referred to the act of measurement as "the act of registration of the result *in the mind of the observer*. The discontinuous change in the probability function . . . takes place with the act of registration, because it is the discontinuous change *in our knowledge* in the instant of registration that has its image in the discontinuous change of the probability function" (our italics).

Or, as Lynne McTaggart expresses in somewhat less scientific terms: "Reality is unset Jell-O. There's a big indeterminate sludge out there that's our potential life. And we, by our very act of involvement, our act of noticing, our observation, we get that Jell-O to set. So we're intrinsic to the whole process of reality. Our involvement creates that reality."

The Foundations of Quantum Mechanics

This area of investigation emerged in the '70s as an attempt to remove the "conscious" part out of the theories of quantum mechanics. It was a much more mechanistic way to view the problem of measurement. In it the physical measurement device was looked at as being the active agent.

As Dr. Albert describes it:

> (There was) a series of progressively more and more embarrassing conversations of the form, "Well, can a cat cause these effects with its consciousness? Can a mouse cause these effects with its consciousness?" Eventually, it was clear that the words involved here were so imprecise, were so slippery, that you weren't going to be able to build a useful scientific theory around them, and the idea was dropped.
>
> This work [Foundations of Quantum Mechanics] has to do with trying to figure out how to alter the equations in order to produce these changes or how to add things, add physical things to our picture of the world, in order to show how these changes come about.

In a nutshell, the Foundations of Quantum Mechanics attempted to look at quantum physics from a purely physical point of view, one that did not include the challenges of a conscious observer.

The Many Worlds Theory

Physicist Hugh Everett proposed that when a quantum measurement is performed, rather than the wave function collapsing into merely one outcome, every possible outcome will actualize. In the process of actualizing, the universe will split into as many versions of itself as needed to accommodate all possible measurement results. This gives rise to the (rather

Einstein's was a universe in which objects possess definite values of all possible physical attributes. Attributes do not hang in limbo, waiting for an experimenter's measurement to bring them into existence. The majority of physicists would say that Einstein was wrong on this point, too. Particle properties, in this majority view, come into being when measurements force them to . . . When they are not being observed . . . particle properties have a nebulous, fuzzy existence characterized solely by a probability that one or another potentiality might be realized.

—Brian Greene,
The Fabric of the Cosmos

unwieldy but definitely mind-expanding) notion that there are innumerable parallel universes where all the quantum potentialities play out.

Take a moment to digest that concept—anytime you make a choice, there are innumerable parallel possibilities or outcomes of that choice occurring *at once!*

Quantum Logic

Mathematician John von Neumann developed a rigorous mathematic basis for quantum theory. In looking at the observer and the observed, he broke the problem into three processes.

Process 1 was the decision by the observer to pose a question to the quantum world. *"Mirror mirror on the wall."* This choice already limits the modes of freedom available to the quantum system in which to respond. (In fact, posing any questions limits the response: If one asked what fruit you had for dinner, steak is not a valid response.)

Process 2 was the evolving state of the wave equation—the process by which the cloud of probability unfolds or evolves in a manner described by Schrödinger's wave equation.

Process 3 was the quantum state responding to the question posed in process 1, which *"collapses the very small."*

One of the interesting things about this formalism was the decision about what to ask the quantum world. Every observation involved a choice about what to observe. Suddenly, words like "choice" and "free will" were being viewed as part of the entire quantum event. Although the question of whether a dog is a conscious observer is debatable, the question as to whether or not a dog ever decided (process 1) to make a quantum measurement regarding the wave nature of electrons seems pretty obvious.

In this quantum logic theory, there is no distinction about

what is included in the physical system involved in process 2. *That means that the brain of the observer could be considered part of the evolving wave function, not just the electrons being observed.* This has given rise to a number of theories about consciousness, the mind and the brain.[1]

Through John von Neumann's quantum logic, a pivotal piece of the measurement problem was brought forward: One decision by the observer makes a measurement. This decision limits the degrees of freedom to which the physical system (such as electrons) can respond, thereby affecting the result (reality).

Neorealism

Neorealism was led by Einstein, who refused to accept any interpretation that concluded that commonsense reality does not exist in its own right, irrespective of our observations and measurements. The neorealists propose that reality consists of the objects familiar to classical physics, and that the paradoxes of quantum mechanics reveal that the theory is incomplete and flawed. This view is also known as the "hidden variable" interpretation of quantum mechanics, which assumes that once we discover all the missing factors, the paradoxes will go away.

Consciousness Creates Reality

This interpretation pushes to the extreme the idea that the conscious act of observation is the key factor in the formation of reality. This provides the act of observation an especially privileged role in collapsing the possible into the actual. Most mainstream physicists regard this interpretation as little more

Neorealistic theory seems to me to say: "We know quantum is wrong because we don't understand it (paradoxes), and we're right because we think so (it's common sense), and we're sure once we know more (hidden variables found) we'll be proved to be right." Sort of like, "We know Elvis is still alive; we just haven't found him yet."

—WILL

---

[1] Henry Stapp, *The Mindful Universe.* This is covered in the "Quantum Brain" chapter.

When we understand the observer, then we have to bow to a greater mind that is forming this energy into modes of reality that we have yet to dream in our lifetime. We only perceive it yet as chaos, but its order is definite. It's above us. It's deeper.

—Ramtha

than wishful New Age thinking and a fuzzy misunderstanding of the measurement problem.

There's a whole chapter in which this is discussed. Suffice it to say that this debate has been ongoing for millennia. The most ancient spiritual and metaphysical traditions have long held the view that, in the words of Amit Goswami: "Consciousness is the ground of all being." Protons and neutrons are relative newcomers to the debate. Their appearance on the witness stand has indeed been a remarkable event.

Wholeness

Einstein's protégé David Bohm maintained that quantum mechanics reveals that reality is an undivided whole in which everything is connected in a deep way, transcending the ordinary limits of space and time. He put forward the concept that there is an "implicate order" from which the "explicate order" (the hidden, non-detectable physical universe) springs forth. It is the enfolding and unfolding of these orders that gives rise to the varieties of the quantum world. Bohm's vision of the nature of reality has given rise to a holographic theory of the universe. This theory has been used by Karl Pribram and others to explain the brain and perception. In a recent conversation with Edgar Mitchell, he felt that the Copenhagen Interpretation is inaccurate, and that quantum holography is a much better model of reality.

Then There's Me . . .

Thus far we have dealt primarily with the physics notion of the observer. The other side of the observer is possibly the most intimate sense that each of us has about ourselves. We have a sense that there is "an observer" somewhere inside us watching, watching all the time. Sometimes referred to as "the still small voice," many spiritual traditions and practices have used the term observer to get a handle on the ineffable self, or to realize our inner nature and by observation change the outer ego self.

The Zen practice of being always present in the moment and not swept away by the external activities could also be described as staying in the observer.

It's no wonder that the impulse to tie the subjective view of the observer with the scientific view is so compelling, especially when the scientific view seems to be talking about exactly that. Subject and object are intimately related. And while our inner sense of the observer is often passive, science seems to be saying that observation is active. There is a physical effect to observation.

And whether or not the "C" word—consciousness—is the sole affecting agent, the fact that any measurement changes the physical system is itself a revelation. It says you cannot take any *information* out of a system without changing the physical-ness of that system.

How Much Does the Observer Affect the Observed?

That's the sixty-four-thousand-dollar question. Says Fred Alan Wolf:

> You're not changing the reality out there. You're not changing chairs and big trucks and bulldozers and rockets taking off—you're not changing those! No! But you're changing how you perceive things, or maybe how you think about things, how you feel about things, how you sense the world.

But why aren't we changing big trucks and bulldozers and ecological demise? According to Dr. Joe Dispenza: "Because we have lost the power of observation." He believes in keeping the quantum physics message very simple: Observation has a direct effect on the observer's world. This will motivate people to focus on becoming better observers. He continues:

We have found that where science has progressed the furthest, the mind has regained from nature that which mind has put into nature. We have found a strange footprint on the shores of the unknown. We have devised profound theories, one after another, to account for its origin. At last, we have succeeded in reconstructing the creature that made the footprint. And lo! It is our own.

—Sir Arthur Eddington

I was always under the impression that I was a pretty cool cucumber. Completely in control of my emotions, my reactions to people, places, things, time and events. Then when I listened to Fred Alan Wolf and to John Hagelin and to our other interviewees, I realized I was nothing more than a marble ricocheting off the walls of life. I'm amazed I ever had time to breathe and didn't suffer a nasty head injury! Becoming more observant of what goes on "inside" and using it to change my perception of the "outside" has opened up the possibilities in my life. Things I never knew I could do or see I can, and time moves much slower and at a pace where I have the ability to observe and choose instead of react and regret.

—BETSY

The subatomic world responds to our observation, but the average person loses their attention span every 6–10 seconds . . . so how can the very large respond to someone who doesn't have the ability to even focus and concentrate? Maybe we're just poor observers. Maybe we haven't mastered the skill of observation and maybe it is a skill . . .

We should be willing to sit down every day and take a piece of our day and set it aside and begin to observe, to design a new possible future for ourselves, and if we do it properly and we observe it properly, we should have opportunities begin to show up in our lives.

Altering your Everyday Reality

Stepping up from the subatomic scale to the human scale: What is observation? For humans, the doorway of observation is perception. Your perception. And remember from previous chapters how suspect that can be? ("Mirror mirror on the wall . . . who's the fairest of them all?") As Amit Goswami observes:

Every observation can be looked upon as a quantum measurement, because quantum measurement produces brain memory. These brain memories are activated every time we encounter and experience again a repeated stimulus. A repeated stimulus will always illicit, not only the original impression, but also this repetition of memory impressions . . .

We always perceive something after reflection in the mirror of memory. It is this reflection in the mirror of memory that gives us that sense of "I-ness," who I am, namely a pattern of habits, a pattern of memories, a pattern of past.

In other words:

Memory (past) → Perception → Observation → (affecting) Reality

Is it any wonder that practices such as "A Course in Miracles" stress *forgiveness* as the important element for changing the present? And think about the teachings of Jesus—how much he taught forgiveness. And about perception: "Before you find fault in your neighbor for the speck in their eye, first remove the log from your eye." And the ultimate observation: "Love your neighbor as yourself."

The subtitle for this book is "Discovering the endless possibilities for altering your everyday reality." Well, if reality is just the answer to the questions, or attitudes, held in the mind, and that answer lays down at the end of a long chain of memories, perceptions and observations, it's not so much how do we alter reality, as much as it's a wonder *why* we keep our reality the same. In the answer to that is the key to change.

The measurement problem is only a "problem" because it so radically undercuts the notion that we are outside of the observed. Even a simple measurement device interacts with the measured system and changes it. There is a fluidity to the observed reality that seems contrary to the world of coffee cups and rockets taking off. And yet it is a fundamental feature of how all aspects of reality hook up to each other.

And the operable words are "hook up." Or we could say connect, or entangle, or are parts of the same wave equation. This notion of the essential inseparability of all things keeps popping out from the quantum witness stand.

And who are we, mere humans, to argue with a gazillion million billion electrons?

Who collapses the very small? Not who—*what.* Every thing.

The question remains—is it just *things,* or also *no-things*: mind, spirit, consciousness? And if they do, are they as real as the things they collapse? In the world of illusions, the separation between things and no-things may be the one illusion that all the rest of the illusions hang on.

"From the quantum perspective—the universe is an extremely interactive place," science writer Dan Winters comments in a Discover article with the provocative title, "Does the Universe Exist if We're Not Looking?" The article summarizes Princeton physicist John Wheeler's idea of "genesis by observership." According to Wheeler (a colleague of Albert Einstein and Niels Bohr and coiner of the term "black hole"): "We are not simply bystanders on a cosmic stage; we are shapers and creators living in a participatory universe."

Ponder These for a While . . .

- Is it possible to identify yourself as the observer if you are the observer?

- What or who is self?

- What or who is the observer?

- Are they separate?

- Can you observe something within yourself other than self?

- If you can become an observer to your "self," how will that change your perception of reality?

- If it takes the observer to create reality, how focused of an observer are you? What reality are you creating in your current state of observing?

- How long can you hold a thought?

- Does reality continue to exist when you are not observing it?

- If the observer is required to collapse reality, then what keeps your body together when you are sleeping?

- Who or what then is the observer?

CONSCIOUSNESS

Consciousness is a very difficult thing to define.
People have been trying to explain consciousness, and trying
to figure out what exactly it is. What it means for us
as human beings. Why we even have it.

ANDREW NEWBERG

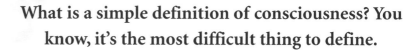

What is a simple definition of consciousness? You know, it's the most difficult thing to define.

—Fred Alan Wolf

Nick Herbert earned his Ph.D. in experimental physics from Stanford University. For years, he was senior scientist at Memorex Corporation in Silicon Valley, and worked in magnetic, electrostatic, optical and thermal physics. He wrote *Quantum Reality: Beyond the New Physics*, and taught science at all levels (from kindergarten to graduate school). Remember Bell's Theorem (proving the existence of non-locality) from previous chapters? Nick Herbert has created the most concise mathematical proof of this revolutionary feature of the universe that demonstrates interconnectedness beyond space and time. And what is he interested in now? *Consciousness.*

"My real notion is that *consciousness is the toughest problem*," he says, "and that physics has basically taken off on the easy problems. . . . We may find all the forces and all the particles of nature—that's physics' quest—but then what? Then we have to really tackle some of these harder problems—the nature of mind, the nature of God, and bigger problems that we don't even know how to ask about yet."

What Is Consciousness?

We all have it. (Don't we?) We are all aware, conscious beings. (On a good day.) We have seen that quantum physics has brushed up against consciousness in its quest for answers about reality and perception. "It" goes with us all the time: Every sense experience, thought, action and interaction plays out on the field of consciousness.

Consciousness is fundamental to all that we do—art, science, relationships, life; it's the constant of our lives. And yet science has done very little to examine it deeply. In its nearly 400-year life span, "science has made immense progress in comprehending the physical universe at all scales from quark to quasar," says Herbert. But consciousness remains "an intellectual black hole."

Many scientists, in physics as well as psychology, who are still wedded to the materialist/Newtonian paradigm, dismiss consciousness as a product of brain functioning. (The word they most often use is *epiphenomenon,* which means essentially a side effect or by-product.) Essentially this says the "me" sense of you is an "oops" accident of evolution. And that when the brain dies, the "oops" goes away, and the packaging joins the other consumed wrappers in the dump.

Barking Up the Wrong Tree

If consciousness is so important and fundamental, why is so little known about it? One explanation is that it's like looking for your glasses while they're sitting on your nose—it's just always there, and so it's taken for granted. Another reason is that we live in an extremely materialistic age, which has been dominated by a materialistic science; in other words, as a culture, we are interested in the "stuff out there," and not so interested in what goes on "in here."

Even when we turn our attention inward, we've been more interested in the *content* of consciousness, the stuff that populates the neuronets—thoughts, dreams, plans, speculations—than in consciousness itself. We're interested in the

Most of us take for granted the word and the idea of consciousness. There is this sense I have when I look down at my body that something is driving it. Something animates it. Mostly, we never ask what that something is, because we are it. The times that this has been most profoundly noticeable are on occasions that I have been out of my body and looking back at it. In that moment I see the form of a human body, which takes me awhile to recognize as myself. In those moments I am something that still exists, that still "thinks" and still has a sense of self. This experience is beyond language, which is maybe why nobody has successfully described what consciousness is using words.

—MARK

images of the movie, but we forget that without the screen on which the images play, nothing would be there.

But probably the most important reason is that *consciousness doesn't fit in the Newtonian paradigm.* It's not made out of measurable stuff. You can't put a meter on consciousness. And most scientists remain immersed in the worldview split apart hundreds of years ago by Descartes: The intangible, or non-physical, or spiritual are forever separate from the physical. Therefore to explain consciousness they have only a brain-based phenomenon of chemistry and neural circuits. And in that paradigm, scientists have gone so far as to call consciousness an *anomaly.*

Say what? My consciousness, your consciousness, the basic fact of our existence, an *anomaly*—a *"deviation from the normal"*?

The fact is that currently science doesn't have the framework to understand consciousness. It's a "hard question"—so for the most part, scientists have turned their backs on it and gone on to other things. This is standard procedure when paradigms are challenged (and people's livelihoods are at stake). "When paradigm anomalies first arise," notes physicist and philosopher Peter Russell, "they are usually overlooked or rejected."

Unraveling the Bouncing Balls

In the late 19th century, the Newtonian model of the world as a giant mechanism filled with solid things that bounced into each other like billiard balls was standard doctrine. As Columbia University physicist David Albert has pointed out, "This sufficed very, very well until around the middle of the 19th century when investigations into electromagnetism by people like Faraday, and culminating in the work of Maxwell at the end of the 19th century, began to make it look as if particles weren't the complete catalog of the furniture of the universe anymore." The electromagnetic phenomena could not be explained in terms of the accepted principles of physics; yet they could not be dismissed. So, against the prevailing winds of orthodoxy, it became necessary to take fields into account. As Dr. Albert says, "Fields had been talked about at least as early as the beginning of the 19th century, but for a long time they were not taken seriously." Now they were accepted as intrinsic, fundamental aspects of the cosmos.

It may be time to do the same with consciousness, and the challenge may be similar. As electromagnetic charge and the electrical and magnetic fields had a different quality or nature from the solid objects that were at the heart of the Newtonian model, consciousness is an even more subtle level of reality than even forces or force fields. But, if it is a level of reality, and if physics is to come up with a genuine "theory of everything," consciousness will have to be included.

It's an interesting point of view, this quantum bandwagon. For example, is consciousness only quantum? Is it only explained by quantum physics? I used to think so, back when I was thinking, back in the seventies. I don't think so anymore. I'm now convinced that, uh, quantum mechanics may not be enough.

—Fred Alan Wolf, Ph.D.

THE CHICKEN OR THE EGG?

We have all been operating, observes Ed Mitchell, within "a scientific model which said that everything can be reduced to just matter or just energy. Everything is really energy, with matter a form of energy. Further, according to this model, consciousness as we experience it is simply epiphenomenal. That is, it's a by-product of brain activity; it's not really fundamental.

"And yet all religious traditions the world over have said in one way or another, 'No, this is not so. Consciousness itself is what is fundamental, and energy-matter is the product of consciousness.' This is the basic issue we have been dealing with for a long, long time without really being able to clearly resolve it one way or another."

The Anomalies

"This profound shift in physicists' conception of the basic nature of their endeavor, and of the meanings of their formulas, was not a frivolous move: It was a last resort."

—Henry Stapp

Remember that line? It was in the "Observer" chapter, and it talked about how the strange behavior of the subatomic world *forced* scientists to shift. Or as it has been said: "It was a beautiful theory destroyed by an ugly fact." Are there any "ugly facts" forcing the scientific world to adopt "last resorts"?

- Most people have had at least one experience that cannot be explained by normal means or the blow-off line: "just a coincidence."
- Scores of near-death experiences have been studied that report very similar experiences *while out of the body.*
- Past-life regressions have produced obscure facts from times past that the subject had no access to and which have turned out to be true.
- Remote viewing—the ability to transcend space and time to gain information—has been so successful that the United States and Soviet Union had teams of remote viewers doing spy work.
- Experiments show statistically how human intent changes random quantum processes (and not just collapsing wave functions either).
- Miraculous healings.
- Prophetic dreams.

The list goes on. It starts looking a little silly (some would say ugly) to insist that all of these data points are just illusions, coincidences and swamp gas. And to deny that consciousness may be a reality unto itself.

As I become more aware of myself and my connection to everything, I realize that I am consciousness or an expression of it. I have a feeling of being connected to something, which is greater than myself or my personality. This inspires me to dig and see what I can manifest from it. I love that feeling I get when I've been quiet even for a moment and received a sign from out of nowhere and I act on it. That's when the magic happens.

—BETSY

Unraveling Consciousness

Many scientific thinkers, such as Amit Goswami, Peter Russell and David Chalmers (director of the Center for Consciousness Studies at the University of Arizona), are arguing strongly for the inclusion of consciousness within the new framework of science. "If the existence of consciousness cannot be derived from physical laws," says Chalmers, "a theory of physics is not a true theory of everything. So a final theory must contain an additional fundamental component. Toward this end, I propose that conscious experience be considered a fundamental feature, irreducible to anything more basic."

Nick Herbert has come to a similar conclusion. "I believe mind is a fundamental process in its own right, as widespread and deeply embedded in nature as light or electricity," he says.

And science, in its inexorable, steady pursuit, is moving in this direction. Stuart Hameroff and Roger Penrose have put forward a theory of consciousness based on the microtubules in the brain. (We'll get into this in the "Quantum Brain" chapter, along with other quantum-based theories.) Consciousness study programs are appearing in universities. Conferences are drawing a very diverse and interested crowd of scientists, scholars and mystics, all grappling with the seemingly simple question: What is consciousness? But that is no different from the traditional scientific disciplines, grappling with the seemingly simple question: What is reality? As with so many things—it's two sides of the same coin.

And Pete Russell thinks it may be time to flip it: "Rather than assuming that consciousness somehow arises from the material world, as most scientists do, we need to consider the alternative worldview put forward by many metaphysical and spiritual traditions in which consciousness is held to be a fundamental component of reality—as fundamental as space, time and matter, perhaps even more so."

I think consciousness is this awareness we have, this inner life we have of experience, which puts us apart from robots or zombies, who might be having complex behavior, going about their business and doing things without having any inner life or experience.
I would say consciousness is a sequence of discrete events. These conscious events, which are actually this particular type of quantum state reduction due to this threshold at the fundamental level (of space/time geometry), occur roughly forty times a second.

—Stuart Hameroff, M.D.

Is Consciousness King?

In order to bring consciousness into science, at least science as it's currently understood, we have to somehow put a meter on consciousness. Just as there are ways to measure or quantize forces, so this consciousness stuff needs to be. It's part of the analytical process.

There are many texts, both ancient (Hebrew and Kabbalistic) and modern (Urantia Book), that have detailed classifications of the celestial beings—angels, archangels, seraphim, cherubim and the rest. My strange idea du jour is that these are classification systems for ranges and scopes of consciousness. Certainly psychology, which is considered a science, uses categories for getting a handle on personality types, and then uses them to analyze all sorts of things. Are realms of consciousness any different?

—WILL

In fact, most spiritual traditions maintain that consciousness is not "a" fundamental component; it is "the" fundamental component. Every*thing* springs from the underlying well of consciousness.

Dr. Hagelin is convinced this is the case:

The very earliest experience, the beginning of the universe you could say, is when pure consciousness, the unified field seeing itself, creates within its essentially unified nature the threefold structure of observer, observed and process of observation. From that, at that deepest level of reality, consciousness creates creation so, yes, there is a very intimate relationship between the observer and the observed. They are ultimately united as one inseparable wholeness at the basis of creation, which is the unified field, which is also our own innermost consciousness, the self.

And with all the unanswered definitions and questions about consciousness, how have Dr. Hagelin, Teresa of Avila, Christ Jesus, Lao Tse, Eckhart Tolle, the Vedantic Sages[1] ever achieved any clarity on this issue? *By using consciousness to investigate consciousness.* Just as the physical scientists use physical measuring devices, the explorers in consciousness use that vehicle to discover that reality. Remember Ed Mitchell's recounting of his moment of cosmic consciousness? It seems a common experience to those who have it—the boundaries of the self go away to reveal that the self is everything, everywhere, all the time. Suggesting that not only is consciousness not created by the brain, but that the brain limits consciousness. (Which is probably a good idea when you're driving.)

And what's it like when the explorers in consciousness return? Says Dr. Andrew Newberg:

When people have that (mystical) experience, they

---

[1] This list could go on for quite some time . . .

perceive it to represent a more fundamental level of reality than our everyday material reality that we normally live in. In fact, even when they're no longer having that mystical experience, they still perceive that reality to be the more real, to represent the more truer form, the more fundamental form of reality. And the material world that we live in is kind of a more secondary reality for them.

If that is the case, then why wait for the shocking, over-the-top mystical experience to wake you from your material reality? What happens when you begin *creating* mystical experiences within your consciousness? The answer, of course, is that you'll change your reality.

But are you changing the reality "out there"? Certainly as the physical world becomes secondary, your experiences are fundamentally changed. For instance, the scratch on the new car isn't as big a deal, *but* can the scratch be changed? The good news and the bad news is—only you can determine that for yourself. It's "bad" because no one can make that discovery for you, "good" because once you do, no one can ever tell you *no*.

If we change our heads about who we are—and can see ourselves as creative, eternal beings creating physical experience, joined at that level of existence we call consciousness—then we start to see and create this world that we live in quite differently.

—Ed Mitchell

Ponder These for a While . . .

- Is consciousness the unified field? What are you basing your answer on?

- Most people who have a near-death experience report the same thing: being "out of their bodies," going down a tunnel, seeing a light at the end of it and feeling bliss. Can these accounts be turned into a science? If so, how?

- How many people have to agree on something before it's a "fact"? What about the Middle Ages when everyone agreed the Earth was flat?

- What other anomalies do you know about? Which ones have you experienced?

- If consciousness is "the ground of all being" and the first cause, how can you ever be "conscious of consciousness"?

- How would you define the relationship between reality and consciousness? Is it hierarchical (one creates the other), or a tangled hierarchy (chicken and egg)?

- Why did the chicken (consciousness) cross the road (reality)?

- Although that's kind of funny, it is a real question: If consciousness is the first cause, why is it compelled to experience reality? Why are you?

MIND OVER MATTER

All that we are is the result of what we have thought.
The mind is everything. What we think, we become.

BUDDHA

Well? Is it true? Does mind over matter actually happen, or is it a delusional experience of schizophrenic malcontents who find life so boring they must manufacture fantasy properties of this very solid world?

Don't you want to know if there is any evidence that this is so? Isn't it a question that everyone really, really wants to know the answer to?[1]

W e know that matter affects mind. Matter over mind. Here's a simple experiment you can do to prove this:

- Note your mental state.
- Place a large piano, upright or grand, three feet above your foot.
- Release.
- Note your mental state.

In this example, unless you're Roger Rabbit, Daffy Duck or Wile E. Coyote your mental state is significantly different. Not surprising because matter is the solid, substantial stuff, while mind is the ephemeral, empty stuff. Right?

---

[1] Sorry, Higgs Particle—you're outvoted.

Scientific Method

The previous exercise was not only a thought experiment you just employed, but also the cornerstone of science—the scientific method. As Dr. Jeffrey Satinover explains:

> The scientific method is the most objective human method of investigation ever. It's absolute; it is not bound to any culture; it is not bound to sex; it is an absolutely powerful tool for the investigation of reality in the hands of anybody who's willing to use it.

Simply stated, the scientific method is this: Take a theory, devise an experiment that tests that theory in a way that all extraneous influences are eliminated, run the experiment and, if the theory is contradicted, look for a new theory.

Let's face it. In this day and age, we look first to science to tell us answers about reality. Mind over matter is a controversial issue in modern science. Another experiment: Ask ten people whether they would like to know if there's any scientific backing to the concept of mind over matter. (Not to mention the fact that if mind over matter is a reality, then people's acceptance of that reality makes it infinitely easier to do it.)

Throwing Down the Gauntlet

We've been talking about paradigms and the natural resistance to change, but in the end science is run by scientists, who are people. At a recent conference, John Hagelin remarked to the attendees, "Don't make the mistake of thinking that scientists are scientific."

Let's not mince words here. In this area of mind over matter, psychic research and paranormal activities, prejudice in the scientific community runs rampant. It is an affront to the very methodology that they expound.

I think the interesting thing about physics is that it is a genuinely new and novel way of trying to come to grips with the world. I think the experimental method, which is important to physics, is a very different business from the method of revelation or the method of meditation.

—David Albert, Ph.D.

There are scientists who are as prejudicial as human beings as anybody else. There is the scientific method, which is specifically a method to minimize the influence of prejudice. That is what science itself is.

—Jeffrey Satinover, M.D.

There's a whole realm of physics called the hidden sector, which is given to us by superstring theory; it's a world of its own. It pervades this space; we walk through it; we can even, in principle, see it dimly. This is probably what we call mind. There's probably thought bodies and thoughts that dwell there like physical creatures dwell here.

—John Hagelin, Ph.D.

Why do we care? Because so much that goes on in our world is based on the scientific understanding of the day. And the history of science tells us something great: Once science truly takes something on, its march is relentless, throwing out theories and conjectures until it finds the ones that fit with the experimental evidence.

Dr. Dean Radin has been running experiments at the Institute of Noetic Sciences for many years and has been at the forefront of the battle to get science to recognize evidence about psychic and mystical phenomenon—essentially, mind over matter:

> I tend to push on the evidence. The evidence is a lot stronger than you think, and there is a lot more of it than you think there is. I push it in the same way that I would push any prejudice; in order to fight a prejudice, whether it is racial or gender or some other prejudice, you have to take an affirmative stance.
>
> So I take an aggressive stance just as I do with affirmative action and say there is something to look at, go look at it. . . . Once you start actually paying attention, in this case to evidence, you realize that everything we look at in an evidential sense is filtered through theory. So if your theory is that it can't exist, then you're not looking at the evidence in the proper way.

The Experiments

As Dr. Radin points out, there is a lot of evidence for mind over matter. One example is Random Event Generator (REG) experiments that focus on intention. These devices (sometimes called Random Number Generators) are essentially an electronic flip of the coin. They are based on either a single quantum event, like radioactive decay, or a composite of many cascading quantum events, typically "noise" that electronic circuits generate.

Dr. Radin tells about his experience with these experiments:

Back in the 1600s, when Francis Bacon was developing empiricism in science, he . . . mentioned dice as a possibility. Every time a die bounces, you can trace all the way down to quantum events that caused it to turn this way versus that way. So if it bounces a bunch of times, it basically becomes a quantum mechanical uncertainty.

When electronics came around, someone had the idea of simulating the die in electronic circuits. The reason that became useful was because it became very easy to measure precisely what was happening, and you could record it automatically.

In other words, once electronics came into play, you could rely on the machine to record your observations, eliminating human error. The result was not just more accurate observation and recording, but an explosion in REG experiments.

One type of random number experiment that has been conducted hundreds of times over the past four decades or so has been a random generator that only produces sequences of random bits, of zeros and ones, like flipping coins. You simply ask somebody to press a button that produces two hundred bits, but you ask them to try to make it produce more one bits than zero bits.

When you take the entire body of literature, all of the hundreds of experiments that have been done, you can ask a single question: *Did it matter* that people were trying to push it toward ones or push it toward zeros? And the overall answer is yes, it does matter. Somehow intention is correlated with the operation or the output of the random number generators. If you wish for more ones, somehow the generators produce more ones.

. . . The final analysis is **fifty thousand to one**. The odds against chance [being the reason the generators went the direction they did, toward intention] are fifty thousand to one.

Every day I wake up and remind myself of my favorite quote of Ramtha's, "The only way I will ever be great to myself is not what I do to my body—but what I do to my mind." Ultimately whatever I do with my mind affects my body because it's all the same thing. It reminds me to get out of body/mind consciousness and into mind=body=reality consciousness.

—BETSY

I've always wondered about this business of "You can only change your perception, not the physical world." Just before I hit puberty, I remember screaming at myself in a mirror. It was probably something to do with hair that wouldn't behave in the way I wanted it to. A few beats after I finished yelling at myself and was staring at my reflection with open hatred, the mirror shattered. It was as though the pieces flung themselves off the surface of the cabinet. I remember standing there in a state of utter shock. To me it was apparent that I had done this with my anger. In the years since, I have ruled out sound vibrations, freak weather anomalies and the statistical chances of the mirror's structural failure coinciding with my outburst of anger. I am left with the resounding realization that my mind or my emotions had done it.

We have heard many stories of children being able to do this until puberty, at which time they lose the ability. What is it about their change in focus that no longer allows them to do this? Does this ability have more to do with a special state than some law about physical reality? Maybe the reason so few adults can do them is because they don't know how to reach that special state. And what if they could?

—MARK

There has been criticism that his findings "are only statistics." But the quantum wave function is just the probability of *statistically* finding a particle at a given place at a given time. So if it is a problem, it's in good company.

Random Event Generators: Collective Mind

Remember the O.J. Simpson trial? How can we forget—hundreds of millions of people awaiting the fateful guilty/not guilty verdict. For those millions it was high courtroom drama. For Dean Radin, Roger Nelson and Dick Shwope, it was a chance to see if not just *intention*, but *coherent minds* could push REGs out of randomness.

What would happen with hundreds of millions of people all had their attention focused suddenly on something? And as it turns out, about a month from the time I had that thought, there was going to be the reading of the O.J. Simpson verdict. This is an unusual point in human history in that people knew far in advance that within a fraction of a second, they were saying the words guilty or not guilty. Something of very high interest would occur which would attract hundreds of millions of people live.

They decided to record the event through random number generators. They set up three in the lab in the United States, one in Amsterdam and one in Princeton. With all five random number generators ready to record the moment the verdict was announced, the scientists waited to see what would happen.

We ran the generators, and afterward we evaluated the results, and sure enough we saw a spike with odds of a thousand to one, actually in two places; one was when the camera switched from outside the courthouse to inside the chambers, which got a huge rise of attention, which was reflected in the random generators; the other was at the moment that the verdict was read. There was this large spike of coherence in all five generators at once.

A spike of coherence refers to a graphical representation of the randomness. Normally the REGs are running 50% 1's and 50% 0's. So the graph of 1's versus 0's is flat. But for some reason, when millions of people focused on the same thing, that flat graph deviated sharply from 50/50 at precisely the dramatic moment of focus. This contradicts a basic premise of quantum theory—that quantum events are purely random.

Since then, Radin and his colleagues have launched the Global Consciousness Project. In this experiment, REGs around the world run twenty-four hours a day, and every five minutes send the results to a server in Princeton. They have seen significant spikes during events like Y2K, 9/11 and the funeral of Princess Diana. The statistics are mounting, and as Bill Tiller says of his experiments: "The results are robust."

Intention Imprinting Electron Devices (IIED)

Bill Tiller was the head of his Material Sciences department at Stanford. But decades ago he decided to forego the department head position, government committees and power positions to focus on "this other stuff." He specifically set out to verify experimentally whether or not human intention affects physical systems. Not "just" collapse a wave function or two, or "just" push a random quantum event around, but affect a macroscopic attribute of matter.

So he constructed an IIED. It is a simple box, with a few diodes, oscillator, E-prom, some resistors and capacitors. Then:

> We set it on a tabletop around which four very well qualified meditators, highly inner-self managed individuals sit, and they go into a deep meditative state; they cleanse the environment; they make it essentially a sacred space using their mind and their intentions. And then one of the four speaks the specific intention for this device.

Emoto's Photographs of Water

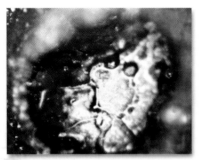

Polluted Water from the Fuiwara Dam

Sanbu-Ichi Yasui Spring Water

Blessed Water from a
Fountain in Lourdes

The intention is to influence a particular target experiment. To increase the pH of purified water by one full pH unit, or to decrease the pH by one full pH unit, or to increase the thermodynamic activity of a specific liver enzyme, alkaline phosphatase. Or to influence an in vivo experiment with fruit fly larvae to increase their energy molecule ratio in their body (the ATP to ADP) so they become more fit and have a shorter larval development time. We have used these devices on all four of those experiments and have been robustly successful.[2]

In addition to the boxes that have been imprinted, "control" devices (no imprint) are prepared. They are separately wrapped in aluminum foil and shipped to a lab a thousand miles away. Then both types of boxes, control and imprinted, are placed about six inches from the target and turned on. It takes three or four months for these boxes to "condition the space to a higher symmetry state." In other words, for them to work. The bottom line: "We see marked contrasts within any pair [and within] all of them [collectively]. We see big effects with statistical probabilities better than one part in a thousand."

Simply put: Dr. Tiller has four meditators focus on a simple electronic box to intend something—like the pH of water changing one full unit. They ship it off, place it next to some water, and a few months later the pH has changed. There is less than a thousand to one chance that this change would have occurred naturally, especially given that the change did not occur with the control units.

How big is one pH unit? Says Tiller: "If the pH in your body changes one full unit, you're dead."

As for how this is accepted, Dr. Tiller notes: "Normal scientists have difficulty with it . . . the boggle effect comes in. Their eyes get a little glazed, and they would prefer not to continue the conversation."

---

[2] For a detailed description, see *Conscious Acts of Creation,* by William Tiller.

Messages from Water

Dr. Masaru Emoto has made a splash with his book *The Hidden Messages in Water.* The book features stunning photographs showing frozen water crystals after they have been subjected to non-physical stimuli. He began by subjecting the water crystals to music—from Beethoven to heavy metal—and photographing the results. When the music clearly affected the size and shape of the water crystals, he moved on to consciousness. Music, after all, creates a physical, material object that can affect matter-sound waves. But what about thoughts?

Dr. Emoto put signs on bottles of water that expressed human emotions and ideas. Some of them were positive, such as "Thank You" and "Love." Others were negative, such as the sign that read "You Make Me Sick, I Will Kill You." Contrary to the prevailing wisdom of science, the water responded to these expressions of consciousness, *even though the words did not create a measurable physical action.* The water with the positive messages formed beautiful crystals; the water with the negative messages became ugly and malformed.

The response to these photographs has been worldwide. Between his books, our movie and Emoto's tireless trips around the globe giving lectures and seminars, there has been a huge public groundswell for more information about his experiments. In response to this, a number of scientific researchers are busy replicating his work. Independent replication is part and parcel of the scientific method.

What unites all of humanity, in fact all of life, is water. Between 70 percent and 90 percent (depending on whose research you read) of our bodies is water. The surface of the planet is mostly water. In a brilliant insight, Dr. Emoto goes right to the heart of the one physical element that *life* has in common. If life (us) can affect the physical, it seems most natural that it would show up in water.

As is evident from all the above, there is much for the scientific community to consider. Experiments have been run, and

**Water Reacts
to Consciousness**

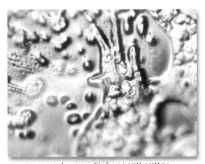

You Make Me Sick, I Will Kill You

Thank You

The Chi of Love

more are continually being run. Results are being published. And, meanwhile, most of the world still really wants to know: How real is mind over matter? If thoughts can do that to water, imagine what they can do to us.

Mind *Over* Matter?

Assuming that mind over matter is a feature of reality, and that we proved, or at least convinced you in our thought experiment with the piano, that matter over mind is also a feature, what does this mean?

Mind over matter over mind over matter over—it's another tangled hierarchy, another chicken and egg aspect of the universe. But as Ramtha notes, that view is intrinsically dualistic. Dualism pervades these concepts: subject/object, in there/out there, science/spirit, consciousness/reality. That worldview that we've been making such a to-do about once more creeps into our language and our thoughts. What about mind as matter and, therefore, matter as mind?

How about matter as information or mind as information?

It's at times like this that the suggestive pull from quantum physics is nearly overwhelming. Does the fact that matter ends up looking like information *prove* the view: mind as matter? Well, it certainly doesn't disprove it; in fact, it seems to *suggest* this view is going in the right direction.

It suggests it like a piano falling on your foot suggests pain. Like observers (conscious or otherwise) affecting the observed, like particles being connected across the universe suggest a non-dual world. They don't suggest a non-dual world; they prove it. The dream from Newton of a divided universe is over, and in this spirit of positive affirmative action everyone wonders: What are we going to do with *that*!?

Ponder These for a While . . .

- What prejudices do you have that keep you from shifting to a new paradigm?

- How do those prejudices reflect in the things in your reality (your stuff)?

- What is "your stuff"?

- Is it possible that by knowing that those things are a manifestation of your thoughts, you could easily manifest new things based on your new paradigm?

- List five things that are different between mind and matter.

- Could you look at those things in a different way and see them as the same?

- If thoughts can affect the molecular structure of water, what are your thoughts doing to your reality?

- Which came first—the chair you're sitting in or the idea of sitting in the chair to read this book?

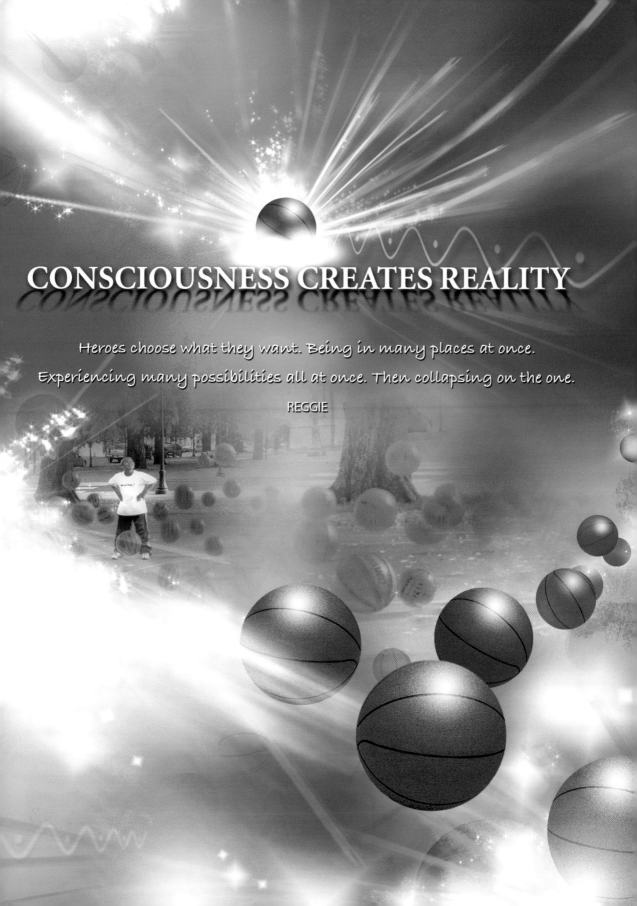

CONSCIOUSNESS CREATES REALITY

Heroes choose what they want. Being in many places at once.
Experiencing many possibilities all at once. Then collapsing on the one.

REGGIE

**All roads lead to Rome
Everything so entangled
Eternal City**

**Eternal Question:
Do I make reality
Or does it make me?**

All the roads we've been trekking in the previous chapters really lead up to this one. And from this eternal question, all the chapters will lead out. None of us can escape without somehow answering: Do I create my reality, or am I a leaf in the wind? Am I the source that determines the things in my life, or is my life at the end of a chain, determined in one instant at the big bang?

We saw in the "What Is Reality?" chapter how we answer the reality question every time we get out of bed, every time we interact with what's "out there." Well, is "Do I create my reality?" answered in all of our internal "in here" moments? And if it's true that we do create our reality, then those "in here" moments are the precursor to those "out there" moments. Which is why this is the central "in here" moment of the book.

This idea has been, and still is, a central concept of spiritual, metaphysical, occult and alchemical traditions. "As above so below, as within so without" is held as a fundamental, true way to see the world. Yet while common sense says you create some events in your life (what to have for breakfast, what person to marry, what car to drive), it seems a stretch to say that you had anything to do with that tree falling on that car.

In fact, the concept of you creating reality (it does after all get created somehow—it's there!) has a myriad of nuances. It generates questions like:

- If I create and you create and they're different—what then?
- "I would never create this (fill in the blank) in my life!"
- Do coincidences exist?
- Does a starving child create that?
- What about natural disasters?
- Who is the "I" that is creating?

And these questions in turn tie into the concepts of karma, transcendent self, frequency specific resonances, attitudes, personal responsibility, victimization and power.

But the bottom line is: Which side of the fence you are on with respect to this concept has the single biggest impact on the life you live.

All speech, action and behavior are fluctuations of consciousness. All life emerges from, and is sustained in, consciousness. The whole universe is the expression of consciousness. The reality of the universe is one unbounded ocean of consciousness in motion.

—Maharishi Mahesh Yogi

Back to the Lab (Again!)

In "Mind over Matter" we saw how intent seems to push around events on the microscopic level. We saw how the supposed randomness of quantum events could be altered, and how the focus of our minds could bring about a different physical state. In "Observer" it was all about collapsing a cloud of fuzzy possibilities into a definite single state. In "Quantum Physics" it was how this solid fixed reality isn't so solid, fixed and stable, and there is a connection between everything in the universe. The parallels between "Quantum Mechanics" and "Consciousness Creates Reality" are once again compelling.

According to Princeton's Nobel Prize–winning physicist John Wheeler, "Useful as it is under everyday circumstances to say that the world exists 'out there' independent of us, that view can no longer be upheld." In Wheeler's words, we are not simply "bystanders on a cosmic stage, [but] shapers and creators living in a participatory universe."

According to physicist and author Amit Goswami, "We all have a habit of thinking that everything around us is already a thing, existing without my input, without my choice." To be true to the findings of quantum physics, Goswami says, we "have to banish that kind of thinking. Instead, you really have to recognize that even the material world around us, the chairs, the tables, the rooms, the carpet, all of these are nothing but possible movements of consciousness. *And I'm choosing moment to moment out of those movements to bring my actual experience into manifestation.*"

What these physicists, and the new physics in general, are spelling out is the death of dualism. It's not mind over matter; it's mind = matter. Not consciousness creates reality, but consciousness = reality.

Think of the two sides of the fence.

consciousness	physical reality
mind	*matter*
spirit	*science*
transcendental self	*nature*
god	*things*

In chapter after chapter, we have been looking at the relationship between the two sides. We've been looking for causal relationships. Who causes what? Is there a connection? Is there a split? Who created the split, and who sits on the fence with a foot in both realms? We do, and we did.

But with the death of dualism, there is no connection or cause (or fence). It's all the same thing. Everything is interdependent—which is what those explorers in consciousness have always maintained. Goswami freely admits how difficult it is to adjust to this new way of thinking that seems to contradict our daily experience. He says, "This is the only radical thinking that you need to do, but it is *so* radical. It's so difficult, because our tendency is that the world is already out there, independent of my experience. It is not. Quantum physics has been so clear about it."

All of this led Fred Alan Wolf in the '70s to coin the term, "I created my reality." The then emerging New Age movement immediately seized upon this phrase and made it part of *their* paradigm. But as many physicists went on to say, it's not a simple concept to really grasp fully. As we quoted Dr. Wolf earlier: "You're not changing the reality out there; you're not changing chairs and big trucks and bulldozers and rockets taking off—you're not changing those!"

Who Is Creating What?

Dr. Wolf continues: "One of the things that comes up about creating realities is what happens when there are two people each creating a different reality; what goes on there? Well, the first thing to realize is that the idea that you create your own reality, if by you, you mean that egotistic person that you think is running your show creates your reality, it's probably wrong. It's probably not that you that's creating the reality at all." Says Amit Goswami:

> It became clear that the place from where I choose to create my own reality, that place of consciousness, is a very special non-ordinary state of being where the subject/objects split and disappear. It's from this non-ordinary state that I choose, and therefore the exultation of the New Ager disappeared when it was forced to face the reality that there really is not a free lunch. We have to meditate and reach non-ordinary states of consciousness before we become the creator of our own reality.

So the concept is, "consciousness creates reality" brings in the questions: "What consciousness? What level of consciousness? Which 'I' is creating?"

A wonderful depiction of this question is the movie *Forbidden Planet.* In it, the people on the planet build a machine that instantly transforms their thoughts into physical reality. The big day comes, and they turn it on and

My son, Evan, the physicist, says it's an additive thing if I'm holding one reality and someone else is holding another reality . . . It's the Super Bowl this afternoon, and the reality that the Eagles are holding, is a different reality than what the Patriots are holding, and only one of those realities is going to be the real reality.

—Candace Pert, Ph.D.

WOW! What a day. They create wonderful mansions, a Ferrari in every driveway, beautiful parks, sumptuous banquets, after which they depart (in the Ferraris) to the magnificent mansions and fall asleep. And dream.

And wake up the next morning to a devastated planet.

According to Dr. Dean Radin, there's a very good reason why we don't manifest things right away: "Everything you do, everything you think, all your plans spread out and affect the universe. As it turns out, though, most of the universe doesn't care, and that's why our individual little thoughts don't immediately go out and change the universe as we see it. I can imagine that if that were the case, if each of us were so powerful that our fleeting whims would go out and affect the universe, we would go out and destroy ourselves almost instantly."

Think of all the times someone cut you off on the highway, and you thought (you know what you thought) and that immediately became reality. Or the time your spouse did _____ and you said _____!

The lack of immediate gratification in creating our own reality could be to protect us from ourselves.

Attitude Is Everything

Two of the fundamental ideas taught by Ramtha at the Ramtha School of Enlightenment are: consciousness and energy create the nature of reality, and attitude is everything. The first idea states the law of *how* things got the way they did, and the second idea is *why*.

A recent study at Harvard by Ellen Langer and Rebecca Levy compared memory loss in elderly people of different cultures. Mainstream Americans, who live in a culture that fears old age and "knows" that as we age our powers decline, had substantial memory loss. By contrast, elderly Chinese, whose culture holds older people in high esteem, not only showed very little memory loss, but the oldest performed almost as well as the youngest people in the study. Each culture produced old people in keeping with the prevailing attitude about aging.

And then there's the French, whose culture is fine with drinking wine, smoking cigarettes, eating pastries (refined sugar!), cholesterol-clogging sauces and living to a ripe old age healthy, thin and happy. Many studies have been done in an attempt to figure out what "the secret" is, because based on the current theories, there should be a coronary bypass for every pastry shop. It's not a secret: It's attitude. They love what they eat and feel no guilt about it.

If it's personal, we call it an attitude—if it's cultural, we call it a paradigm—if it's universal, we call it a law. *"As within, so without."*

As Above So Below

Remember how we said the "What Is Reality?" chapter could go anywhere? Well, this chapter *is* everywhere. "Sight and Perception" talks about how the brain creates the images that we think of as the real out there world. "Paradigm Shift" looks at how ideas and discoveries bubbled up into a widespread belief about the world. "Quantum Physics" and "Observer" offer tantalizing parallels between the subatomic world and how consciousness interacts with the creation. "Mind over Matter" kicks down the fence between the seen and the unseen and points to a connection between these two seemingly different realms. In other words, they're all about consciousness, reality and the relationship between them.

Going to the future, the "Brain 101" chapter is about how our attitudes are encoded in the neural structures, and what gets created out of that. "Emotions" and "Addictions" are answering the questions about "why I create the reality I find myself in." "Why Aren't We Magicians?" looks at why we don't create what we (think) we want. "Desire/Intent" looks at how to consciously wield the axe of creation, while "Choices/Changes"

> There was an experiment done, where people were given the task of eating something decadent. And the people who did it with a sense of guilt and shame actually experienced a transient decrease in immune function. Whereas those who were able to just enjoy the experience and savor this wonderful whatever that they put in their mouths actually had a surge in immune status.
>
> —Daniel Monti, M.D.

> I can't really prove that you are out there in reality and have your own consciousness, and you can't prove that I'm conscious.
>
> —Stuart Hameroff, M.D.

There were some big cases of applying Consciousness Creates Reality—like entertainment professionals all saying no one will distribute it, no one will see it—but there were hundreds of "little" things we did. We actually counted on this to get us out of jams. Like when we had to change a song. To transition to the party scene, we had the Aerosmith song "Sweet Emotion." In fact, the scene had been written with that in mind. But we couldn't get it. So we searched and searched for something else, but nothing was even close.

One day I was thinking about it and got frustrated and said (quite intently), "I'm gonna hear the song on the radio, NOW!" "Hooked on a Feeling" by Boston came on. Not only was it great, but when I cut it in, it actually fit the scene better! Is this proof? Was to me.

—WILL

(in the same chapter) looks downstream of the axe swing to see what happens in our life.

But in the cosmic court of experience, does any or all of the above prove that consciousness creates reality? The concept seems to be reflected throughout the different levels of experience ("As above so below"), but is that proof or circumstantial evidence? We've heard from everything, from electrons and positrons to distinguished physicists to filmmakers. According to Dean Radin: "*Prove* is not a word that is used in science. We can show evidence for it. We can create a certain degree of confidence that an effect is what it appears to be. Has anyone ever 'proved' gravity? Newton said gravity is the force of attraction between masses. Einstein said mass curves space/time geometry, which then causes masses to come together. But they can't prove that's what it is. At best the mathematical description can be seen to have no evidence to the contrary."

It's at this point that we filmmakers threw up our hands and leaped off the fence.

Creating *What the BLEEP Do We Know!?*

We took a vow. We weren't going to be hypocrites. We weren't going to be armchair philosophers. To the best of our ability, we were going to not just talk the talk, but walk the walk. That meant we were going to live the ideas in the film, and of all the ideas, the two biggees were: emotional addictions and create your own reality. During the making of *What the BLEEP Do We Know!?*, those involved tried to really work to create reality and tackle emotional addictions.

In other words, we decided to be "the scientists in our own lives" by employing the scientific method, by trying it out to see if it worked. And the result was thumbs-up, with a better way to live life.

But that's just three of the creators. The movie was entitled *"What the Fuck Do We Know?!"*[1] so it's not clear you should listen to us. Who should you listen to? The person that in reality you always do—you.

MASS TO MASS

I was directing the scene in the Baghdad theater. We had to shoot from midnight 'til dawn, and we had to be out of there by then. We were way behind. I realized we were never going to "make our day," which was a disaster because then the entire scene wouldn't work. Ramtha is always getting on our case about working "mass to mass." In other words, all our creating is by pushing physical things physically, instead of creating on a subtle level and letting it go.

I realized we were screwed, and I had no way out going mass to mass. I remember thinking, "It's time to create like a God" and running up those ramps in the theater shouting, "No more mass to mass." That's about all I remember after that, except that by dawn we had gotten it all done.

—MARK

---

[1] There! We said it! That is the real title after all . . .

Ponder These for a While . . .

- Take the two supposed sides of the fence:

consciousness	physical reality
mind	matter
spirit	science
transcendental self	nature
god	things

  Draw lines between all of them specifying what the relationship is. Then draw lines with the simple relationship "is" (mind is matter, reality is consciousness). Which view makes more sense?

- Are they mutually exclusive?

- Are these two views yet another creeping influence of dualism? Is sex an attempt to end dualism?

- Was time invented to keep instant karma at bay?

- Was it invented to give us *time* to realize our power and the ramifications of it?

I CREATE MY REALITY?

I live the life I love and I love the life I live.

WILLIE DIXON

The previous chapter was about considering whether Consciousness Creates Reality is a law of the universe, or a fantasy, whether there was any supporting evidence that it is in fact true, and how that evidence may appear in our lives.

So by now you've somewhat made up your mind about where you stand with the idea that you "create your reality." Everyone admits that it's true to some degree. The question is, how far do you take it? As far as whether or not to go to the ice-cream store, or as far as believing that the leaf that falls on your head was a creation of yours?

The implications of this principle are huge. Not just to us and the life we live, but also to bigger lives: cities, states, countries and the planet. But first—what about you?

What's for Breakfast, What's for Life?

You'd probably agree that in countless small ways, you create your life every day. You decide whether or not you're going to get up when the alarm clock goes off. You decide what to wear, what to eat for breakfast, or maybe to skip breakfast. And as you run into people throughout the day, at home, work or the freeway, you decide how to treat each one. Your intentions for the day—or your default decision not to set any intentions but just to float on through—affect what you do and what you'll experience.

In the bigger picture, the whole trajectory of your life is generated by your choices. Do you want to get married? Have children? Go to college? What to study? What career? What job offer to accept? Your life doesn't just "happen"; it's based on the choices you make—or don't make—every day.

But the question remains, How far does that life-making extend? As far as that chance meeting with the girl of your dreams? As far as the tyrannical boss? Winning the lottery?[1] And who's life are YOU making anyway? Seems like a dumb question, but the "I" in "I create my reality" is a big question mark. And answering it gives some clarity to this whole creation stew.

We're reality-producing machines. We create the effects of reality all the time. If we take information from a small knowledge base, we have a small reality. If we have a large knowledge base, we have a large reality.

—Joe Dispenza

Who Am I?

Back to great questions. The Indian sage Ramana Maharshi built his teachings around this very question. According to him, the inquiry into this question leads directly to enlightenment. But let's postpone enlightenment for now and limit the question to the act of creating . . .

According to Fred Alan Wolf: "The first thing to realize is that the idea that *you* create your own reality, if by *you* you mean that egotistic person that you think is running your show creates your reality, it's probably wrong. It's probably not that you that's creating the reality at all." But this begs the question: "So who is?" Certainly when you order that first cup of coffee in the morning, it's pretty clear that the "egotistic person," or the personality decided upon the double cappuccino, and not the transcendental, immortal self. And that when the tree lands on your shiny new car, the personality had nothing whatsoever to do with that.

Most often people reject the "I Create Reality" idea when

---

[1] Actually, everyone takes credit for that one. "After what I've been through, I had it coming."

We are running the holodeck. It has such flexibility that anything you can imagine, it will create for you. And you learn. Your intention causes this thing to materialize once you're conscious enough, and you learn how to use your intentionality.

—William Tiller, Ph.D.

something occurs in their life that they absolutely, positively would never, ever create. "I would never create *this*!" That's true; they—the personality—never would. But as all spiritual traditions maintain, there's more than one "you."

This divine schizophrenia goes by many labels: ego/true self, personality/divinity, son of man/son of god, mortal body/immortal soul, but in essence it says there are different levels from which you are creating. And the goal of enlightenment is to erase this fragmentation of self and create from one source. (Which I guess is why "Who Am I?" works.) It's to expand our consciousness until we are conscious of all our creations.

And accepting that "I create . . ." is an amazing tool for that expansion. For if it's true, then every time you reject your part in creating reality, you are rejecting or denying a part of yourself. Thus the fragmentation continues. In fact, according to the Enlightened, the spirit half of you is creating these realities for the sole purpose of becoming whole. There are things you must experience to grow that might not be your ego/personality's first choice.

They call it karma: We did create, at some time in the past, whether recent or remote, all the conditions that we are faced with in this life. But how do all the karmas of all the people in the world interact? How does it all fit together? How do those happy (and unhappy) "coincidences" occur that are often the harbingers of a new world? Who is running the computer that can keep all of this straight, for 6 billion humans?

How Does It Work?

The universe IS the computer. Non-duality. And it doesn't have to run. It is connected, entangled so that it is hooked into everything and is created from everything. It doesn't respond to us—it *is* us.

The dualistic model of karma says: I hit Bob, so someone

will hit me. It's a very cause → effect (a.k.a. Newtonian) way to view this phenomena. But from the non-dualistic, entangled model, it's different. It says that action or thought (which are the same "thing") arises in a piece of my consciousness. There is a certain frequency or vibration associated with that. By taking the action, I endorse that reality so that I am now connected to the universe by that frequency or vibration. Everything "out there" of the same frequency will respond to it,[2] and they will then be reflected in your reality.

By this notion, everything in your life—people, places, things, times and events—are nothing but reflections of your signature vibrations. According to Ramtha: "Everything in your life is *frequency specific* to who you are." So if you want to know "Who Am I?" just look around; the universe is always serving up the answer.

The trouble is the hidden, repressed parts of us are also reflected, and we repress them because we don't like them. It's those reflections that make us say: "I would never create *that*." And it's *that* which keeps getting reflected back over and over until we understand it. That's the wheel of karma. The un-merry-go-round. Or as a high-school philosopher once said, "Life's a shit sandwich, and every day you take a bite."

Said like a true victim.

"Gravity doesn't exist; the earth sucks." Ditto.

"Life's a bitch, then you die."

Victimization—Cure for the Current Reality

Perceiving one's self as a victim is possibly the strongest rejection of "I create my reality." And it happens all the time. The victim says: "This situation *happened to me*. It is unfair and unwarranted." The corollaries of this are: "Poor me. The

> *Unto the pure all things are pure; but unto them that are defiled and unbelieving is nothing pure; but even their mind and conscience is defiled.*
>
> —Titus 1:15

I realized recently that the very act of denying our participation as the causal agent in our life is voluntarily trying to use foggy thinking so that we don't have to deal with some aspect of reality staring back at us. I think I perfected this "deception" for much of my life. We always have to wonder what the criteria is when we say we did or didn't create some aspect of reality. My criteria was always: I created what was pleasurable and comfortable, and I tried to deny my responsibility for what was uncomfortable. I got away with it for a while . . . but then reality caught up with me. It always does. I now think that I am the causal and/or participatory force in every aspect of what I see in my life.

—MARK

---

[2] This is the principle that all transmission/reception works on. The transmitter and receiver are tuned to the same frequency.

Universe is unjust. Karma is a part-time, fickle operation."

The upside of this attitude is: You get sympathy, get to feel good about yourself because it's not you, and can blow off the experience and not deal with your part in it.

The downside is: You just endorsed the idea that you don't create your reality (and thus are disempowered to do so), and will get the lesson again and again, and . . . It also is a fragmentation from reality. It removes the Creator from the Creation.

A look at the reflection of this attitude in society at large shows how prevalent victimization is. So much of nightly news is centered on victims. In the United States, the victim mentality has reached epic proportions. If anything happens to someone, the first thing they do is look for someone to sue.

As Don Juan told Carlos Castaneda in *Journey to Ixtlan*: "You have been complaining all your life because you don't assume responsibility for your decisions . . . look at me. I have no doubts or remorse. Everything I do is my decision and my responsibility."

The Big Turnaround

As victimization is the strongest rejection of this chapter's premise, "I accept responsibility" is the strongest acceptance of it. It is a monumental turnaround in the way anyone approaches the world and their experiences in it. Victimization and the powerlessness that ensues is gone from life. In every situation, the questions are asked, "Where am I, or what is I, in this situation? What is being reflected back to me? What level of 'I-ness' is this coming from?"

The turnaround is, instead of asking the universe to prove that you create reality so that you can sit on the fence and accept or reject what happens, you take it as a given that your life and its happenings are created by you, so therefore you look for the meaning in them. And by meaning, it's not a philosophical, cosmic meaning, but rather what does this mean about who you

> We are creating our own reality every day, though we find that very hard to accept—there's nothing more exquisitely pleasant than to blame somebody else for the way we are. It's her fault or his fault; it's the system; it's God; it's my parents . . . Whatever way we observe the world around us is what comes back to us, and the reason why my life, for instance, is so lacking in joy and happiness and fulfillment is because my focus is lacking in those same things exactly.
>
> —Miceal Ledwith

are, or what you're creating, or what in your life you are denying? Looking for change in your life? Make this switch and watch it transform before your many "I's."

"People are always blaming their circumstances for what they are," said the great British playwright, George Bernard Shaw. "I don't believe in circumstances. The people who get on in this world are the people who get up and look for the circumstances that they want, and if they can't find them, make them."

How do we *make* circumstances? How do we make those coincidences that have a huge effect on the direction of our life? It seems crazy that anyone could make a coincidence like: "Well, I forgot the paper, so I had to run back home, but on the way I got a flat tire. So I stopped to fix it and bent over and my pants split. So I wrapped a blanket around myself, and this person drives by, and it was a blanket that she had designed, so she stopped and then we got married." It was just a coincidence. But what we really mean is it was a co–incident.

So did our happy husband create the flat tire? Or did he create being married, and the universe worked out the details? (These are the sorts of questions that come up once you get on the "I Create" bandwagon.) In experiments on creating the pH change in water, William Tiller says: "The issue of whether we do a detailed statement of intention, or if we do it in such a way that we leave it open for the universe to find a way to do it? Generally, it's the latter."

In other words, instead of dictating all the steps the water has to go through to change its pH, such as the rearrangement of chemical bonds, ion exchange, etc. . . . . what the meditators in Dr. Tiller's experiments do is focus on the result, and let the universe supply details, split pants and all.

Possibilities and Time

Still the question remains—how could this all work? And how can someone become more conscious of the possibilities,

As I worked through the many implications of "I create my reality," one of the big ones was: Is my creation a fixed plate or à la carte? Do I choose each and every occurrence, or are there "package deals"? I thought about Jesus and the Passion. Did he create the suffering and abuse, or did he set out to bring a new awareness to the world, and the abuse was just part of the deal?

—WILL

When the syllable "co" is before something, it means some type of interrelationship. Cooperate means to operate together. So coincident means the elements of the incident have an interrelationship. Strange that the word now means the opposite.

so that creation itself is more conscious? According to Amit Goswami:

> It is a hypothesis that consciousness is the ground of being. It's all possibilities of consciousness. Out of these possibilities of itself, consciousness chooses the actual experience that it manifests, that it observes . . . quantum is talking about possibilities, but when you look at yourself, how many times have you wondered "what possibilities"? Your wondering of possibilities . . . [might be] confined to such trivial things like what kind of ice cream would I choose this time, vanilla or chocolate, which depends on your past experiences totally. So you miss the quantum physics of your life.

Dr. Goswami sees the possibilities in one's life spread out like the probability waves of an electron. This means those options in your life are as "real" as those waves predicted by the Schrödinger equation. Stuart Hameroff takes this concept one step further:

> Each conscious thought can be thought of as a choice, a quantum superposition collapsing to one choice. So let's say you're looking at a menu, and you're trying to decide whether to have shrimp, pasta, tuna fish. Imagine that you have a quantum superposition of all these possibilities coexisting simultaneously. Maybe even you go into the future a little bit and taste the different meals. And then you decide, "Ah-ha. I'll have spaghetti."

Going into the future is not as science fiction as it sounds. As Dr. Hameroff points out: "In quantum theory, you can also go backward in time, and there's some suggestion that processes in the brain related to consciousness project backward in time."

If all these theories prove to be correct, that means that an individual's consciousness is constantly scanning all the future possibilities, maybe even going into the future to "taste" whether to marry someone or not, and then focusing, or collapsing that chosen possibility into reality. The "how" gets handled by the immensely interactive superintelligent universe that automatically responds to consciousness because that's what it is. The universe IS the computer that keeps track— that's why it's here. And if it can create self-replicating, self-conscious life forms, it can fix a flat tire.

And how does this view make creation more conscious? Well, to many people the future is on the other side of a great wall, past which they cannot go. So those possibilities lurking out there are not seen, and when they do show up it's a surprise, or a shock. But realizing that those potentials are real, and they can be developed, manipulated and collapsed, takes us over the wall and into the future where the new you awaits.

Creating Your Day

Your pool of created reality lies out in front of you. Smeared across the landscape of time, those possibilities await "the movement of consciousness" to bring the actual event into experience. But let's say you're a bit more proactive—a landscape activist who isn't willing to sit back and let the weeds of the universe happen to you, but rather seed that landscape of possibilities with your conscious creations.

The most popular, captivating, requested information in *What the BLEEP* was the concept of creating your day. This technique was first taught by Ramtha in 1992 to his students and is one of the cornerstones of the school in Yelm, Washington: "No masters worth their salt ever let the day happen to them; rather, they create their day."

ONE LITTLE TEACHING HAS
BECOME THE RAGE

This all really sounds cool in theory. But it's the doing it that's tough. I remember the first few times I began to create my day. In that moment before I knew who "I" was, it was impossible. My whole body would go into shock. I would get frightened and have to do something to bring myself back to "reality." It was as if I couldn't stand to be away from all of my normal identity long enough to create something different. Fear and panic took over. It took a long time to just allow myself to be afraid but to continue into the unknown, and as I began to see the effects and results, the fear turned into anticipation.

—BETSY

The following is from the DVD *Ramtha: Create Your Day—An Invitation to Open Your Mind:* "You know the moment you wake up, have you ever noticed, by the way, that you don't know who you are, and you wake up and you don't know who you are. Have you noticed how you look around the room to orient yourself, and what is really surprising is when you see the person next to you and for that split moment you don't know who they are? I think you should contemplate that a lot. We spend the next moments before you ever get out of bed reorienting, rebonding with an identity that for a moment we didn't even have, and the identity is that [thing] we start to form when we take a look at the person next to us. And then we get up, and we start scratching our bodies . . . and then we get up and we go to the latrine and on the way we look at ourselves. Why do you do that? Why do you stare at yourself? Because you are trying to remember who you are. It is still a mystery.

"But if you have to remember who you are and remember the parameters of your acceptance and the fence of your doubt, if you have to go through the ritual every single day to remember who you are, what are the chances that your day is going to turn out unique? . . . Very slim indeed. But what if . . . before you tried to remember who you were that you remembered what *you wanted to be,* and maybe that came first before you saw your mate, before you clawed yourself, before you staggered out of bed, scared the cat and saw yourself in the mirror. Before you did all of that you remembered something: 'Before I bond to the ritual of my neuronet, I am going to create a day that is astounding, that will add to my neuronet, that will add to the experience of my life,' and you create your day—create your day. In that moment that you are not yet who you are is the most sublime moment in which in that moment you see the extraordinary, you can expect and accept the unordinary, you can accept a pay raise today. If you become yourself, your expectation of a pay raise greatly diminishes. You and I both know that. But in this one state of nonconclusiveness about your identity, you can create anyway.

"So I tell my students, before you get up and remember who you are, create your day. Then after you create your day, your routine will change. You will be a slightly different person staring at the urinal, looking in the mirror. There will be something different about you, and that will be a wonderful thing."

This marvelous teaching addresses the "I" that has ultimately been the subject of this chapter. Who is the "I" that is creating? If it's the personality, then the creations are from the existing structures, habits, propensities, neuronets, and from that old personality structure, all that will be created is the same old, same old. Creating what has already been is hardly creating.

Or creating is coming from the higher self, the God self, in which case it's usually unconscious and the workings of some deeply buried karma. So while the creations are wonderful to the spirit, to the disconnected personality, they seem arbitrary, unfair, and give rise to the feelings of powerlessness and victimization.

Whereas this technique takes advantage of the moment of no-self, or, new-self. From this "I" something truly new can be manifested. Something that you *consciously* create. And to create this way forever undoes the trap of victimization and disempowerment.

And it affirms every day, in a very real way, that you create your reality.

And if *that* is true, the affirmation is gas on the fire.

Ponder These for a While . . . .

- What are the limits, if any, to our creativity and power?

- Can we change the laws of physics? If so, are they laws? What are laws?

- Learning to create more effectively, what kind of responsibility do we have?

- What is the most constructive use of our creativity?

- How can we know that our individual aims are aligned with cosmic aims?

- What is the impact of knowing that we are creating all the time, whether consciously or not?

- What is the difference between the personality and the higher level of consciousness?

- How do I know the difference?

- When do I know my personality is creating, or when do I know it's my higher consciousness?

- Is my personality bad?

WHY AREN'T WE MAGICIANS?

LUKE SKYWALKER, UPON SEEING YODA LIFT HIS X-WING FIGHTER OUT

OF THE SWAMP USING ONLY HIS MIND: I don't believe it!

YODA: And that is why you fail.

Magicians.
Leaping off the written page and silver screen . . .
Master Yoda, Headmaster Dumbledore,
Gandalf the White. Wizards! Wielders of
the fire of creation—Majick!

Who wouldn't want to be a magician?
Who wouldn't want to transfer
from Middle School #405 to Hogwarts?
Who would really rather be a Muggle?[1]
Who? Obviously—Muggles!

For IF we create our reality, and there's no magic in it, then we created *that,* didn't we? It does seem strange that most people love the idea of magic and would love to be adept at it, yet the TV remote is as close as we get.

Which brings up an interesting point: What is magic as opposed to science? Certainly a TV remote (and the TV!) would be high magic 200 years ago. The ancient Sumerians wrote about the gods who could magically communicate by voice instantaneously around the planet. And today it's as simple as deciding on the 50-minute or 100-minute plan. So what's the difference? Science is seen as very procedural, but so is magic: Books on magic have step-by-step procedures, like a cookbook, on how to produce a desired result. It seems then

---

[1] Muggle, from the Harry Potter books, refers to someone with no magic. Hogwarts is the school, run by Dumbledore, where magic is learned.

that magic is simply science on the other side of the paradigm. And what Dr. Tiller is doing with his black boxes is pushing that boundary.

The UnWizard's Handbook

Instead of worrying about why we're not magicians (since only you can answer that), we're going to take a few pages from a chapter in *The UnWizard's Handbook,* entitled "How to Turn a Magician into a Toad." And if some of this starts sounding familiar, that will give a clue as to who else has been reading, and using, this handbook.

#1. Convince people that they are NOT magicians.
#2. Teach the glories of becoming a Victim.
#3. Confound and crosswire Belief Systems.
#4. Make New Knowledge scary and inaccessible.
#5. Make Magicians creepy and being a Magician dangerous.
#6. Get them to Lie.
#7. And never look inside.

And just in case you happen to be a Wizard who got turned into a toad, *The UnWizard's Handbook* also has antidotes to each of the above.

#1. Convince people that they are NOT magicians.

Since everyone is a magician, if you convince them they are not, then that's what they'll be. In that case, read no further.

Antidote: Remember your greatness: You are already a Magician.

If you are told that you are a happenstance mutation of

"It was an age of magic, when magic hung heavy in the air. The trees breathed their song to birds that told tales of power. Enchanted glens held secrets that the proper word would call forth gold to shimmer out of nothing. Magic was everywhere. And I was there. Immersed in the scintillating splinters of creation that I called real."

The preceding was from a book I'll never write. But how many of us have read those books with a longing for a teleporter to take us there? In an age of magic, it is everywhere. Because everyone KNOWS that's the way the world is. Before Westerners showed up in Tibet, the stories of "magic" were commonplace. The Lamas had an ability to hop in ten-meter hops so that they could cover hundreds of kilometers in an hour. Commonplace, and gone today. Which then begs the question: "What do we want for tomorrow?"

Maybe I will write that book after all . . .

—WILL

matter, and that your sense of you is a *epiphenomenon* property of dead particles, how magical do you feel? If your magical moments are explained away as coincidences, where's the magic in your life? If your view of the world is death, disease, war and suffering (a.k.a. evening news), all of which you feel powerless to do anything about, how much of a magician are you?

What about being told that you are a despised sinner, born in sin, and wretched in the eyes of God?

About forty years ago, a group of pioneering thinkers, led by Brandeis University professor Abraham Maslow, became acutely aware that psychology seemed almost entirely preoccupied with *problems and disease*: neuroses, psychoses, dysfunctions. Why not study healthy or even supremely healthy individuals? Why not look at what the higher and highest possibilities are for human beings—and create ways to help everyone unfold their powers?

The great American writer, Henry David Thoreau, had written a hundred years earlier, "Man's capacities have never been measured; nor are we to judge of what he can do by any precedents, so little has been tried."

Norman Cousins, author and editor of the *Saturday Review,* expressed the same idea: "The human brain is a mirror to infinity. There is no limit to its range, scope or creative growth. No one knows what great leaps of achievement may be within reach of the species once the full potentiality of the mind is developed." This is what the Human Potential psychologists set out to discover.

Perhaps the greatest legacy of Maslow and his peers is that they helped make it common knowledge that everyone has tremendous latent potential, that there are powers and capabilities within all of us that never come to the surface, yet!

#2. Teach the glories of becoming a Victim.

Once magicians accept being a victim, they have relinquished their claim to creating reality. For victims, reality happens to them: It is unfair and never their fault. So they never have to look within where they will see their own creation.

Antidote: Accept responsibility for your life.

The glories of being a victim are many. You are never "at fault," so you don't have to feel guilt. People feel sorry for you and pay attention to you and help you. And blame is the name of the game. Blame parents, society, job, partners, health and blah blah blah for the circumstances in our lives that we're not happy about. But this excuse network comes crashing down if we accept that *consciousness creates reality* (CCR). This is the most practical aspect of CCR. It means that *you* have created *your* life and *your* world. You may bitch and moan because you can't seem to have what you want, when in fact, you *do* have what you want. You are living the life you chose to live, *the life you believed you could live.*

To really know where your consciousness is truly at, simply observe every thing, every person, every place and every event in your life.

> Accepting that "I create my reality" wasn't easy or fun. I looked around at the carnage and chaos I had created and thought, "Shit, this is a mess!" But hey, if I can create that, then I can create something else.
>
> **—BETSY**

#3. Confound and crosswire Belief Systems.

Belief is the engine of creation. Any glitch in the belief of a magical act will derail it. "Authorities" are very useful in this undertaking.

Antidote: Don't give power away to authorities and trust your own experience. Remember: Belief is the engine of creation.

Did Jesus walk on water?

Do thousands every year walk on a hot bed of coals and not get burned?

Betsy's husband Gordie firewalked three times without ever even feeling it. Then he asked his physiology professor about it.

He was told that it really didn't happen, and that the red-hot coals weren't really hot. It was the "Leidenfrost" effect. On walk number four, Gordie started wondering if what the professor said was true, and whether it really was hot. He came off the coals with third-degree burns.

So what did he do? First, he applied antidote #2—instead of blaming the professor or the firewalk coach, he realized how *he created that outcome,* and then applied antidote #3. He didn't buy into the Leidenfrost effect, but into his own experience. Walk number five was no problem at all.

It's a tricky thing, this bit about belief. Who wants to be a crazy wacko, imagining happenings so far from reality that the disconnect will drive you crazy? But then again who wants to be so limited by the average, run-of-the-mill, day-to-day reality that there's never any magic anywhere? The idea of walking on water is crazy. If it wasn't for that famous scene on the sea of Galilee, everyone would tell you, without a doubt, that it's impossible, and if you think it's possible—you're nuts!

It's a tricky thing because our belief systems usually are not an integrated, well-thought-out, consistent whole. In other words, there may be a moment of exhilaration where you believe you can fly, but you still believe that what goes up must come down. According to Miceal Ledwith:

> If I believe that I can walk on water, and I scream and shout to myself I'm going to walk on water, usually what I'm emphasizing is a disguise on the doubt that I have, so if you go out with that divided focus in your mind to walk on water, you're going to sink.
>
> But if you had absolutely and fully accepted it, yes, it would happen because reality accommodates precisely to our intent. Likewise, if you can focus and thoroughly accept that you're healed, it will happen, but *what we don't really accept is the degree of doubt* that hides behind those affirmations.

And if you dig down far enough, doubt is always the result

of a belief system, which collides with the belief that you're acting upon. And until those conflicts are eliminated, the X-wing fighter stays in the swamp. As Ramtha says:

How broad, how deep, how high is the level of your acceptance? Because that is what belief is. You can never, ever manifest in your life that which you do not accept. You only manifest that which you accept. So how broad is your acceptance? Is it greater than your doubt? What are the limitations of your acceptance? Is that why you are sick? Is that why you are old? Is that why you are unhappy, because the level of your acceptance is unhappiness? That is all you get, you know. You don't get anything greater.

#4. Make New Knowledge scary and inaccessible

New knowledge is the key that unlocks old belief systems and opens the door to greater and greater realities. Furthermore, knowledge strengthens one's belief in the true workings of the universe, thus empowering the Magician. Therefore, employ the most drastic of countermeasures: fear.

Antidote: "Seek and ye shall find, knock and it shall be opened unto you."—Matthew 7:8

When we think of a Magician, visions of someone in his study, with books lined to the ceiling and scientific instruments strewn about, come to mind. Dr. Dispenza explains why: "You have to have knowledge that allows you to dream greater dreams. You have to have the willingness and the passion to step outside the boundaries of your own comfort zone and your own need for conformity, to try doing something that humanity would consider insane. And yet, done without the proper knowledge, it is insanity. Done with the proper knowledge, it's considered heroism. It's considered brilliance. It's considered genius."

Why is Donald Trump rich? Because he's accepted making $3 million a day. He has built a model of possibility that most other people don't have. Most other people have a model of maybe they can make $200 a day. He's just added a bunch of zeros and accepts it. He wasn't born with some special gift necessarily. It's all about acceptance. How do you build your level of acceptance? Athletes build by training, and then by running a race. They find out what their limitations are. They find out what they are good at. Their coach helps them, and they eventually may win Olympic medals. Why should it be any different in looking at the mind? Shouldn't we be training the mind, too? Shouldn't we be giving the mind information?

—MARK

Amit Goswami concurs: "Whenever our boundaries are expanded, we feel happy. Whenever our boundaries are contracted, and we become identified with a small cocoon, we become unhappy. So, the whole idea is to extend our boundaries."

What else did Wizards have in common? Staffs. Both Gandalf and Yoda had one.

"And commanded them that they should take nothing for their journey, save a staff only; no scrip, no bread, no money in their purse."—Mark 6:8

Why a staff? It was considered the staff of knowledge. It supported the worker of magic and guided him on the journey. It also wielded a great power. That is why over the centuries, if you want to turn immortal beings into toads, you burn books. According to John Hagelin:

> Knowledge is the greatest motivator. If people simply know what the potential of life is, what higher states of consciousness beyond waking, dreaming and sleeping offer to life, and what the world will be like, when people are living unity-consciousness, people will be motivated. The only reason people are not motivated is we're not exposed to knowledge.
>
> The media today, what sort of knowledge does the media present to people? Only as much as they need to present to get them to go out and buy a hamburger. Not enough, really. Not enough to inspire people to let them know what's possible in life.

He [Gandalf] stood up and leaned no longer on his staff . . . :
"The wise speak only of what they know, Gríma son of Gálmód. A witless worm you have become. Therefore be silent, and keep your forked tongue behind your teeth. I have not passed through fire and death to bandy crooked words with a serving-man till the lightning falls."
He raised his staff.
There was a roll of lightning . . .
There was a flash as if lightning had cloven the roof. Then all was silent. Wormtongue sprawled on his face.

—The Lord of the Rings

#5. Make Magicians creepy and being a Magician dangerous.

Enlightened beings are magnetically radiant, and they seek to enlighten everyone. Removing them one way or another eliminates the problem and makes others weary of following in their footsteps.

Antidote: Find those that you can learn wisdom from and study.

Bringing enlightenment to the planet has been a hazardous occupation for millennia. Not just for spiritual teachers, either. Scientists also have had a rough go of it. But now that it's more difficult to simply kill off the troublesome teacher, they are made to appear weird, or scary, or creepy. If that doesn't work, the "C" word—cult leader—is thrust upon them. An abundance of people are out there to learn from, and the only way to tell is to see for yourself. Remember antidote #3—don't believe authorities; trust your own experience.

As for it being dangerous, it is, but not as bad as it used to be. Times have changed. Unfortunately, the memory in the collective consciousness remains. The greatest example of this is not the thousands of teachers who were eliminated, but the estimated 50,000 to 100,000 witches who were killed over a 500-year period. It is sad that all the examples of magicians are men, especially when according to many teachers, women are innately more adept at the art. According to Carlos Castaneda's teacher, the Indian Brujo Don Juan, when men are presented with a new magical idea, they think about it and talk about it and debate about it, whereas women simply do it.

How do we take an ordinary day and make it super extraordinary? By having the knowledge to think about the extraordinary. The day becomes it.

—JZ Knight

#6. Get them to Lie.

A lie is a disconnect with reality. It fragments the liars, thus making them a house divided. It also fractures the integrity of their belief systems, thereby rendering any magic they do petty. Therefore, make lying acceptable.

Antidote: On the verge of a lie, ask yourself: What's the worst that can happen if I tell the truth, and is that worth sacrificing my magical heritage?

Today it seems everyone lies. Bosses lie, reporters lie, politicians lie, lovers lie, religious leaders lie. It seems to be acceptable because no one makes much of a fuss about it.[3] But true magicians know the consequences. One of the vows of a Buddhist monk is to never lie. It has been said that because Jesus never ever lied, and he would absolutely never do so, that when he said something, because he never lied, the universe *had* to comply. So when he said, "Arise, take up thy bed, and go unto thine house," it was a law of the universe. "And he arose, and departed to his house." Dr. Ledwith adds:

> "If we had no time for lies in our life or cheating, if we had the absolute impeccability of the spiritual master, then we would do those (miraculous) things effortlessly. But, you know, this is not gained by sitting listening to wonderful music and burning incense and all that stuff. It's attained in a very practical way: by confronting the lies and the untruth and the victimized attitudes that make up what we are. If we root those out, these other things will follow straightaway. If we never root them out, and cover them over with a façade of spirituality, then science and religion can unite 'til the cows come home, and it will not make one whit of difference.

> But seek ye first the kingdom of God and his righteousness; and all these things shall be added unto you.
>
> —Matthew 6:33

#7. And never look inside.

Although the last rule, it is the keystone to all the above. If people never look inside, they will never discover the truth about who and what they really are. Therefore, convince them that true happiness is out there.

---

[3] Case in point: The leaders in America produced a series of lies to manipulate the country into a war that killed tens of thousands of people. And, by and large, America just shrugs.

Antidote: Don't listen—look within.

The daily bombardment by advertisers seeks to stress the notion that fulfillment is gained by attaining something in the exterior world. But the simple fact is, at the end of more cars and houses and money and fame, you are left with yourself.

However, it's important to remember that the advertisers, and the liars and the media are not responsible for anyone's toad-like status. That is being a victim. In the end, it is always each individual who decides what to pay attention to. What to focus on. What to let in. That is our personal choice every time. To blame anything outside is to diminish inside.

Imagine that: walking around with the Kingdom of God, Heaven, Enlightenment, Nirvana *within* you. I mean, literally *imagine* that:

Hold that image in your mind—walking to work, to school, to the grocery store with all *that* inside you. Everywhere you go, you take the eternal with you . . .

If that's not the image of a Magician, I don't know what is.

Behold, the kingdom of God is within you.

—Luke 17:21

WHY AREN'T WE MAGICIANS?

A Matrix of Words

We have been discussing and investigating a lot of concepts. It's time to start tying them together . . .

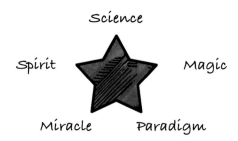

Science

Spirit   Magic

Miracle   Paradigm

Write out what you think is the relationship between these words. For instance, "Magic is outside the paradigm." Go through all the relationships.

As Above So Below

The ancient symbol for this is the six-point star of two intersecting triangles, also known as the star of David.

From the following list of words, see how many above/below relationships you can construct. After you've done this one, feel free to create your own.

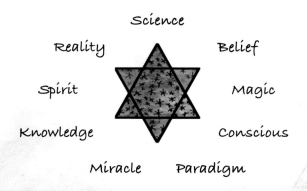

Science

Reality   Belief

Spirit   Magic

Knowledge   Conscious

Miracle   Paradigm

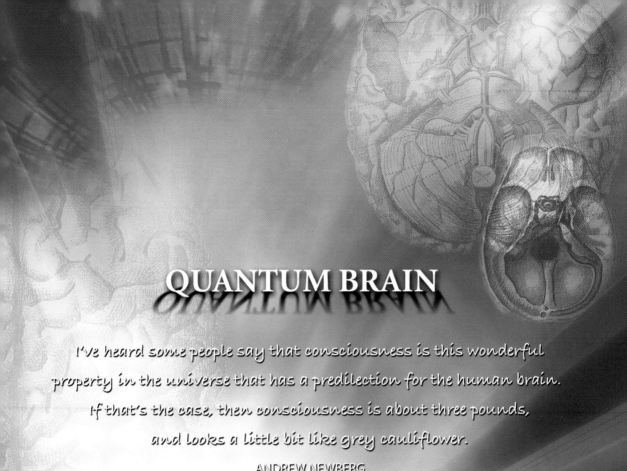

QUANTUM BRAIN

I've heard some people say that consciousness is this wonderful
property in the universe that has a predilection for the human brain.
If that's the case, then consciousness is about three pounds,
and looks a little bit like grey cauliflower.

ANDREW NEWBERG

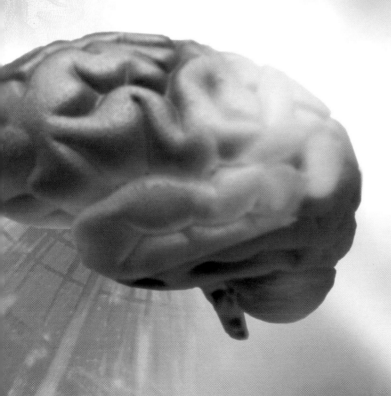

Jeff: "Nowadays you can tag quantum onto anything. A few years back it was creative: creative divorce, creative cooking. Now it's quantum: quantum divorce, quantum cooking, quantum healing . . ."

Betsy: "What is quantum cooking (laughs)?"

Jeff: "I don't know, but it sounds good, right?"

Is it any wonder that quantum is being applied everywhere? It is such a different way to view the physical universe upon which the rest is built. It seems to open the door to so many possibilities and give some answers to puzzles that have stumped humanity for millennia.

One of the most interesting applications of quantum, outside the kitchen, that is, is looking at some of those intangible aspects of our lives: consciousness (that again!), free will, intent, experience and, specifically for this chapter, quantum meets the brain!

Consciousness, the Brain, the Body

Is consciousness merely a product of the brain, an "epiphenomenon" or "emergent property" of bioelectrical activity inside our heads? Is it something that arises when enough neurons fire together, producing a sufficient level of computational complexity? If so, is the brain just a biological computer? And how are we different from machines? Could there be artificial intelligence that equals or surpasses human intelligence? Would such machines be "conscious"? Could they learn? Would they have free will?

Or is consciousness a fundamental component of the universe, independent of the brain, that can be experienced without the body, as in the thousands of documented out-of-body and near-death experiences? In those cases, a person's body temporarily shuts down and ceases to function (on the operating table, for example) and yet their consciousness continues to remain awake to experience.

Historically, the answers to these questions fall into one of three buckets:

The difference between us and a rock is that human behavior as opposed to the behavior of a rock has its roots deep in the quantum mechanical level, from the individual DNA molecules in every single cell in the brain.

—John Hagelin, Ph.D.

- **Materialism:** Matter is primary; consciousness, whatever it is, is secondary. Consciousness is simply an effect of brain activity. There is no such thing as "consciousness" itself; it has no reality of its own, but is merely a product of our biology, of neural nets and electrochemical interactions.
- **Dualism:** Consciousness and matter are both existing realities. However, they are so different (one solid and tangible, the other abstract and intangible) that they operate in entirely distinct, unrelated realms. Descartes, in the 1600s, divided the world into *res cogitans* and *res extensa*—the realm of spirit and thought (*cogitans*) vs. the realm of matter and things (*res extensa*). The material world, including minerals, plants, animals and human beings, is all machines, governed by the absolute laws of causality. There can be no reciprocal action between the free-roaming and abstract realm of pure thought, and the dense and localized realm of matter: They are two utterly different substances.
- **Idealism:** Consciousness is the fundamental reality. Everything is an expression of consciousness. Alive, fluid and perpetually self-renewing, it self-expresses in a continuum of levels or layers, from the most "soft" and abstract pure consciousness through all the subtle and more "substantial" levels (quantum wave functions and particles, photons, atoms, molecules, cells, etc.) to the most solid

matter. In this continuum, everything is connected and related; it is all the same stuff, manifesting in differing frequencies, vibratory levels or densities.

In both the first and second cases, consciousness is either afforded no identity or dignity or is eternally cut off from interacting with the material world. In the third view, the problem of the relationship of consciousness and the body dissolves: They are already and always related and connected; in fact, they are just two aspects of the same "thing." This framework, although considered "extreme" by many, is not only consistent with Buddhism, the Indian Vedic tradition, and the mystics of Christianity, Judaism and Islam, but it is also favored by a number of physicists we have mentioned before. And consciousness could well be the "neutral monism" of the mathematician and philosopher, Bertrand Russell, in which a common underlying entity gives rise to both physical and mental qualities.

But what are the mechanisms by which consciousness loses its pure abstractness, becomes thought, perception and feeling, and appears as electrical or chemical activity in the brain? Here are a couple of theoretical attempts to explain how this works.

Warning: These views are not accepted by the majority of mainstream scientists. Reading these may cause the brain to rewire. Please consult your higher self as to the validity.

Stuart Hameroff's Vision of the Quantum Brain

"How does living matter produce subjective thoughts, feelings and emotions?" asks Hameroff, professor emeritus of the departments of Anesthesiology and Psychology and director of the Center for Consciousness Studies at the University of Arizona. "How can our brains account for phenomenal 'experience'—the smell of jasmine, the redness of a rose or the joy of love?"

Though questions like these have occupied philosophers and thinkers for centuries, Hameroff points out that "the study of consciousness fell upon hard times through most of the 20th century, as behaviorists dominated psychology. Why study something that cannot be measured? Consciousness became a dirty word in scientific circles, eclipsed by operant conditioning paradigms, Pavlovian reflexes and other quantifiable parameters."

A resurgence of interest in consciousness began in the 1970s. Not only were large numbers of people, especially of the '60s generation, actively exploring the transformation of consciousness through meditation, various forms of therapy and an assortment of mind-bending chemicals, but computers made it possible to begin intensive work on artificial intelligence (AI) and to rapidly analyze data derived from electrical readings of the brain (EEG and other measures).

In the 1980s and '90s, numerous prominent scientists jumped on the bandwagon and produced books and theories championing the brain as a magnificent computer, as well as the view that, as Hameroff states it, "consciousness had something to do with the mysteries of quantum mechanics."

Then Hameroff came in contact with the work of Sir Roger Penrose, the renowned British mathematician and physicist.

Penrose-Hameroff "OR" Theory of Consciousness

Penrose proposed that consciousness comes about when superpositions of neurons within the brain reach a certain threshold and then spontaneously collapse. (This is similar to the collapse of the wave function due to observation, bringing a vast array of possibilities down to a localized point value. The difference is that here the superposition collapses of its own accord due to quantum gravity effects.) According to Penrose, what he called "objective reductions" are intrinsic to the way consciousness operates. These "ORs" convert multiple

I began to look at these microtubules and how they might be processing information, and their structure seemed to suggest to me that they were some kind of computer, some kind of computational devices.

Because the walls of the microtubules are very interesting hexagonal lattices with some beautiful mathematical symmetry that seemed to be well-suited for computational operations.

—Stuart Hameroff, M.D.

possibilities at the preconscious, unconscious or subconscious level to definite perceptions or choices on the conscious level, like considering pizza, sushi or pad thai (all in superposition) and then selecting one (collapse or reduction). Hameroff suggested the *mechanism* by which this could take place, and together he and Penrose formulated their theory.

Central to the way this collapse, or OR, takes place are tiny *microtubules*, hollow, strawlike structures within every cell, including neurons. Once thought of as merely the *cytoskeleton* or scaffolding of the cell, microtubules were found to display extraordinary intelligence and self-organizing ability. They serve as the *cell's* nervous and circulatory system, transport materials and organize the cell's shape and motion. They interact with their "neighbors" to process and communicate information, and can organize neighboring cells into a unified, coherent whole. In neurons, microtubules also set up and regulate synaptic connections and are involved in the release of neurotransmitters. As Dr. Hameroff says, "They are everywhere, and seem to organize almost everything."

The structural changes and information processing and communication among the microtubules, within the neurons of the brain, are directly influential "one level up," on the organization of neurons into the networks called "neuronets." But the microtubules themselves are affected from deep within their own structure by a quantum phenomenon: The proteins of which they are made respond to signals from an internal quantum computer consisting of single electrons. Dr. Hameroff explains: "These quantum mechanical forces in the pockets inside proteins control the conformational shape of the protein. And that, in turn, controls the actions of the neurons, and the muscles, and our behavior. So, proteins changing their shape is the amplification point between the quantum world and our affecting the classical world in everything that mankind does, good and bad."

Hameroff continues that it's the spontaneous collapse (OR) of these microtubules, roughly forty times a second, that gives "a moment of consciousness." Our consciousness is not continuous, but a sequence of "ah-ha moments." He says, "Consciousness kind of ratchets through space time, and that consciousness is a sequence of now moments: now, now, now . . ."

Where Does Consciousness Happen?

What is the meeting point between the intangible realm of thought and awareness that constitutes our internal subjective experience and the electrically charged biochemical soup[1] of the brain?

"I'm not an idealist, like Bishop Berkeley or Hindu approaches," says Dr. Hameroff, "in which consciousness is all there is. Nor am I a 'Copenhagenist' in which consciousness causes collapse, and chooses reality from a number of possibilities. But somewhere in between. Consciousness exists on the edge between the quantum and classical worlds. I think more like a quantum Buddhist, in that there is a universal proto-conscious mind which we access, and which can influence us. But it actually exists at the funda-mental level of the universe, at the Planck scale."

The Planck scale has not been introduced, but it is an important aspect of the Penrose–Hameroff theory. The Planck scale (after the quantum physicist Max Planck) is the smallest distance that can be defined. At 10-33 centimeters, it is 10 trillion trillion times (no, that's not a typo) smaller than a hydrogen atom! According to Hameroff, this fundamental level of the universe . . .

> is a vast storehouse of truth, ethical and aesthetic values, and precursors of conscious experience, ready to influence our every conscious perception and choice. We are connected to the universe, and entangled with everyone else

Is the brain quantum? I have no idea. But I can't find any other way to explain how I know something is about to happen or that someone is thinking of me or that they are aware I'm doing the same. It would have to mean that something inside my head is connecting to a super-highway of information independent of time and space. Sounds suspiciously quantum to me.

—MARK

---

[1] There it is: quantum cooking.

through this omniscient omnipresence, a sea of feelings and subjectivity. If we are mindful and not acting reflexively or rashly, our choices can be divinely guided. Penrose avoids any spiritual implication of his ideas, but they are inescapable. Quantum computations in our brains connect our consciousness to the "funda-mental" universe.

Free Will and Chinese Boxes

Satinover, like Stuart Hameroff, a physician steeped in quantum mechanics, has written a book called *The Quantum Brain: The Search for Freedom and the Next Generation of Man*. And while he is reluctant to slap the "Q" word (quantum) on the culinary arts, Satinover has come up with a rigorous mathematical argument that shows that "the workings of the nervous system and the particular way that it implements quantum effects—very distinct, specific ways, not general, not fuzzy, not imprecise—absolutely does open the door for free will being a possibility that does not violate modern scientific tenets." Satinover's theme is directly related to the layered structure described above. It is that the non-determinism of the quantum level of existence, the randomness and the fact that *probability* rather than absolute certainty governs quantum reality, gives us the only possibility for free will.

On the macro level, the large-scale level of classical physics, all events, from the orbits of planets to the movements of molecules, are mechanical and determined by precise mathematical laws. Thus, it is only if the randomness of the quantum level could somehow be relevant on the macro level that choice and free will might be possible.

"At the level of the brain," says Satinover, the neural networks "produce a global intelligence that's associated with the brain as a whole. But then when you look at individual neurons, the interior of the neurons is a different physical implementation of the same principle. And, in fact, at every scale as you go down, like

> There is no place for true randomness in deterministic classical dynamics, and without some source of randomness there are no options. . . . The only known source of perfect freedom of action resides in the quantum nature of matter.
>
> —Jeffrey Satinover, M.D.

Chinese boxes nested the one in the other, each individual processing element at one scale can be shown to be composed of innumerable smaller processing elements within it."

Starting at the "lowest" or smallest scale, the process whereby proteins fold—the process Stuart Hameroff described as operating within the microtubule—"obeys essentially the same mathematically self-organizing dynamic as how a neural network processes information. So, the folding of a protein is mathematically identical to the generation of a thought, or the solving of a problem. And that's really where the notion of the quantum brain comes in. Not that the brain as a whole entity is a quantum entity, but rather that quantum effects at the lowest level are not only capable of, but of necessity are amplified upward because of this nested Chinese box arrangement of the nervous system. . . . It's through a very particular kind of neighbor-to-neighbor interaction amongst the neurons that global intelligence, at the level of the brain as a whole, emerges."

According to this theory, the brain has in fact been designed to magnify these quantum effects and project them "upward" to larger and larger processing elements, until it reaches the level of the brain.

In short, says Satinover, "Quantum mechanics allows for the intangible phenomenon of freedom to be woven into human nature. . . . The entire operation of the human brain is underpinned by quantum uncertainty." This is because "at every scale, from the cortex down to individual proteins," the brain "functions as a parallel processor. . . . These processes form a nested hierarchy, an entire parallel computer at one scale being but a processing element in the next larger one."

Ultimately, what we'd like to [determine] is the physics of consciousness. What is consciousness? Where does it come from? What are the origins of consciousness? What are the limits of human potential? We're in a position to actually answer that now, I believe, although there's certainly not a consensus yet in the scientific community about that. You've asked the questions in the first movie; now we're on the verge of being able to answer those questions.

—John Hagelin, Ph.D.

Intention and Quantum Zeno

The investigations into mind/matter centered, at least on the mind side, around the role of intention—that act, allowed by free will, that chooses the effect that is to be observed in the

The Zeno Effect is that
old paradox that if the rabbit
chasing the turtle closes by
half every time the distance
between it and the turtle,
it will never reach
the turtle.

outside world. And although there is evidence that intention is the key, the *how* that key is turned is up for grabs.

Henry Stapp, a theoretical physicist, has brought the mathematical formalism of von Neumann's Theory of Quantum Mechanics into this area of study. In von Neumann's theory, there are 3 processes in observation (see "Observer"). The first is posing the question. This is where it gets interesting. . . .

According to Dr. Stapp:

> An important feature of the dynamical rules of quantum theory is this: Suppose a process 1 event that leads to a "Yes" outcome is followed by a rapid sequence of very similar process 1 events. That is, suppose a sequence of very similar intentional acts is performed, and that the events in this sequence occur in very rapid succession.
>
> Then the dynamical rules of quantum theory entail that the sequence of outcomes will, with high probability, all be "Yes": The "Yes" state will, with high probability, be held approximately in place by the rapid succession of intentional acts. *By virtue of the quantum laws of motion, a strong intention, manifested by the high rapidity of the similar intentional acts, will tend to hold in place the associated template for action.*
>
> The timings of the process 1 actions are controlled by the "free choices" on the part of the agent. If we add to the von Neumann rules the assumption that the rapidity of these similar process 1 actions can be increased by mental effort, then we obtain, as a rigorous mathematical consequence of the basic dynamical laws of quantum mechanics described by von Neumann, a potentially powerful effect of mental effort on brain activity!
>
> This "holding-in-place" effect is called the Quantum Zeno Effect. This appellation was used by the physicists E.C.G. Sudarshan and R. Misra.

What this says is that by continually holding the same intention over and over and over, by posing the same question to the universe over and over and over, we change the quantum probability away from randomness. Is this what happened when 100 million people were holding the question guilty/not guilty for O.J.'s trial? Or when one person is holding "more 1's than 0's"?

But Dr. Stapp thinks that this phenomenon may show how the insubstantial "mind" controls the very substantial brain: "Quantum mechanics contains a specific mechanism that in principle allows mental effort to hold at bay strong forces arising from the mechanical side of nature, and allows mental intent to influence brain processes."

Quantum Cooking After All

So far we have the experience of consciousness in the brain arising from spontaneous collapse of the wave function in the microtubules. Within this consciousness exists the option of free will, or choice, due to the amplified upward Chinese box arrangement of quantum events. Based on exercising that free will, we hold a given outcome in our brains and re-pose that question arbitrarily fast (which seems continuous like "holding," but is really a sequence of now moments) to effect the probabilities of the quantum world.

Surely that is a veritable stew of quantum ideas, all thrown together to produce a slice of reality.

Or, from the Quantum Cookbook:

- Take a few trillion microtubules, allow to self collapse into an (objective) reduction sauce.
- As quantum uncertainty bubbles up from the bottom of the (brain) pan, separate out one bubble from the array of (possible) bubbles.
- Repeatedly hold this bubble over the flame of consciousness until baked solid (collapsed reality).
- Perceive to taste. Oops, that's taste to perceive. . . .

So, if you wanted to look at it poetically, you would say that human beings seem to be designed to maximize the freedom that's available in their material structure to a degree which mimics the creation of the universe itself.

—Jeffrey Satinover, M.D.

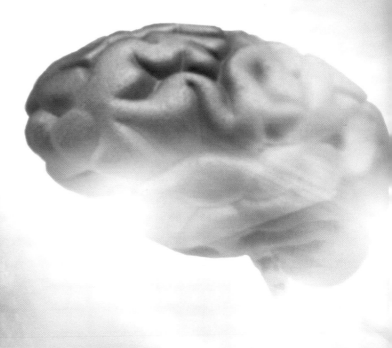

Ponder These for a While . . .

- Given the notion of questions (process 1) being an important element in collapse (process 3), does this explain anything to you about the importance of Great Questions?

- Why do you think we keep asking so many?

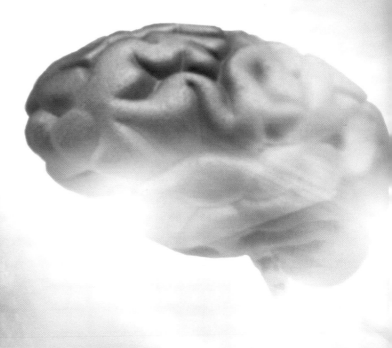

BRAIN 101

The brain acts as a laboratory. It's an architect.

It designs models, and it puts the pieces together.

JOE DISPENZA

It seems funny that humans have explored the ocean's bottom and planetary moons, and developed all sorts of amazing technology, but still are in a mystery about the brain. Scientists are pushed into bringing in quantum effects, complexity theory and holographic models into their theoretical models in order to explain basics like perception, consciousness and memory.

It's not surprising. It has been calculated that there are more possible connections in one human brain than there are atoms in the entire universe. Even in a small brain, the workings are incredible. It's been estimated that to solve the problem of a bird landing on a branch in the wind, the largest supercomputer would take days to calculate a solution, *if* it could. This problem may be computationally unsolvable. Yet bird brains do it all the time, and in no time.

Traditional models compare the brain to a telephone switch-board or a supercomputer. But these comparisons conjure up images of something clunky and machinelike, and the brain isn't like that; it's a very alive, plastic and flexible organ, capable of learning, understanding and dynamically rewiring itself based on our demands.

Even though science is far from fathoming the full extent of the brain's capabilities, there are many things that are known. We do know that it is the most complex structure on the planet, hence in our known universe. It directs and regulates all of our bodies' activities, from heart rate, temperature, digestion and sexual functioning, to learning, memory and emotions. And even though we don't know a lot about how it works, what we do know answers many questions as to why we do what we do.

In the words of brain researcher Andrew Newberg:

> The brain is capable of millions of different things, and people really should learn how incredible they are, and how incredible their minds actually are. Not only do we have this unbelievable thing within our heads that can do so many things for us and can help us learn, but it can change and adapt, and it can make us into something better than what we are. It can help us to transcend ourselves.
>
> And there may be some way that it can actually take us to a higher level of our existence, where we can understand the world and our relationship to things and people in a deeper way, and we can ultimately make more meaning for ourselves and our world. There's a spiritual part of our brain; it's a part that we all can have access to; it's something that we can all do.

What follows is an extremely simplified version of brain structure and processing. The study of the brain is an immensely fascinating area that is great fun to explore. What is discussed here are just the basics that are needed to see how these structures interact with our day-to-day experience of the world and ourselves. We are loading up the tool chest of transformation with knowledge and ideas about how we're wired and unwired, strung up and strung out.

Amazing Facts About Our Brain

- The brain is at least 1,000 times faster than the fastest supercomputer in the world.
- The brain contains as many neurons as there are stars in the Milky Way—about 100 billion.
- Number of synapses in cerebral cortex = 60 trillion.

- A sand-grain-sized piece of a brain contains 100,000 neurons and a billion synapses.
- The brain is always "on"—it never turns off or even rests throughout our entire life.
- The brain continually rewires itself throughout life.

Neurons and Neuronets

The brain is made up of approximately 100 billion tiny nerve cells called *neurons.* Each neuron has between 1,000 and 10,000 *synapses,* or places where they connect with other neurons. These neurons use the connections to form networks among themselves. These integrated or connected nerve cells form what are called neural networks or *neuronets.* A simple way to think about this is that every neuronet represents a thought, a memory, a skill, a piece of information, etc.

However, these neuronets do not stand alone. They are all interconnected, and it's the interconnection of them that builds up complex ideas, memories and emotions. For example, the neuronet for "apple" is not one simple network of neurons. It is a much larger network that connects to other networks, such as the neuronets for "red," "fruit," "round," "yummy" and so on. This neuronet is itself connected to many other networks, so that when you see an apple, the visual cortex, which is also connected in, fires that network in order to give you the image of an apple.

Everyone has their own collection of experiences and skills, represented in the neuronets in their brains. As Dr. Joe Dispenza comments: "Whether we grew up in a single-parent family, whether we were raised with many other children, whether we went to college, what our religious beliefs were, what our culture was, where we lived, whether we were loved and encouraged as a child, or perhaps were abused physically— all of that forms the neuronets in our brains."

All those experiences shape, neurologically, the fabric of what's taking place in our perception and in our world, says

Every time I feel the need to make a change in a pattern/habit, I sit down and visualize the brain and then the neuronets. I find the neuronets that are connected to my particular habit, and I see them pulling away, disappearing. I watch my brain rewire to something new.

—BETSY

Dr. Dispenza, and when stimuli come in from our environment, "certain aspects of those neuronets are going to kick in, or click on, that are going to cause chemical changes in the brain." These chemical changes in turn produce emotional reactions, color our perceptions, and condition the responses we make to the people and events in our lives.

Nerves That Fire Together, Wire Together

A fundamental rule of neuroscience is that *nerve cells that fire together, wire together*. If you do something once, a loose collection of neurons will form a network in response, but if you don't repeat the behavior, it will not "carve a track" in the brain. When something is practiced over and over again, those nerve cells develop a stronger and stronger connection, and it gets easier and easier to fire that network.

If you keep hitting the repeat button in the neuronets, those habits become increasingly hardwired in the brain and are difficult to change. As a connection is used over and over, it gets stronger, better established, like forging a path through tall grass by walking it again and again. This can be advantageous—it's called learning—but it also can make it difficult to change an unwanted behavior pattern.

Luckily there's a flip side: *Nerve cells that don't fire together, no longer wire together.* They lose their long-term relationship. Every time we interrupt the habitual mental or physical process reflected in a neuronal network, the nerve cells and groups of cells that are connected to each other start breaking down their relationship. Dr. Dispenza compares this to an experience most of us have had. When you graduate from college and part with the roommate with whom you've shared so much, you promise to exchange postcards, once a month or so, just to maintain your friendship and let each other know how you're doing. As time passes, you start sending cards only at Christmastime, and the relationship begins to weaken and fade.

If we do something over and over and over again, by the mere fact that we're repeating it, the process of learning whatever we're learning starts to become simple, and it starts to become automatic. It starts to become familiar. It starts to become easy. It starts to become natural, and it starts to become subconscious.

—Joe Dispenza

This effect is an exact reflection of what is going on inside the brain. As you think less and less about the roommate, the neuronets' connection lessens, until there's no connection at all. What's happened is the very fine *dendrites* spreading out from the cell body that connect to other cells unhook and are available to rehook to other nerve cells, letting the old patterns go and potentially forming new ones.

Learning

The brain learns in two major ways. The first way is through factual, intellectual data that we master and/or memorize. For example, you study history and commit names and dates to memory, or you read Plato and draw conclusions about his concept of ideal government. Each name and date, each logical argument, adds to the neuronets in the brain. The more you go over the material, the more firmly established it will be in your memory—because the neural networks become stronger.

The second and generally more powerful way the brain learns is through experience. You can read a book about how to ride a bicycle, and by intellectually processing all the information about how to shift gears going uphill or downhill, how to balance, how tight the spokes should be, you'll get some idea about riding. But it's not until you actually get on the bike and start pedaling that the information is integrated.

Regardless of which type of method, learning is essentially integrating neuronets together to form new neuronets. In the example of the apple, it wasn't just a single neuronet of "apple," rather apple linked into those other neuronets of round, red and so on. Really learning is building new structures based on previous structures. Watch a baby, and you'll see those basic concepts being formed, generally by experience.

Remember the **A Matrix of Words** section in "Why Aren't We Magicians?" (See you formed a net on it!) In it, there was the exercise of linking together a series of words in different relational ways. Each of the links gave meaning to the concepts linked together. And the more the different relational links were examined, the better each concept was understood. That is how the brain learns and how it's wired. And that is why reexamining core ideas and beliefs is life-changing. By reexamining, you can go through all the links and find buried assumptions that are triggering our reactions through the process called Associative Memory.

A HALF-BRAINED STORY FROM JEFFREY SATINOVER

There's a famous story about a resident in radiology in the early days of CAT scans at the Baylor College of Medicine who had a CAT scan of his head, not because he had a problem but just because they were learning about CAT scans. And it turned out that he was missing one half—an entire *half of his brain.* It's a rare condition, but it occurs . . .

And he was a medical resident and as good a doctor as anybody else. It may be an opportunity for lawyers to make jokes about doctors, but the fact is that he didn't need his whole brain, but that doesn't mean that therefore people can or should operate with only half a brain. It's because the functions are distributed over whatever brain is available. And that's in the nature of a neural network to operate that way.

I once wrote down every thought I had for about an hour. I was stunned and amazed at how one thought led to another in this long string of seemingly unrelated ideas. *Oh, I need to call Barry today—you know, I like his restaurant, his chef makes that great steak salad—and then one night that really cute and crazy actress came into the restaurant with dreadlocks— reminds me of that other weird hand model I worked with in South Africa fifteen years ago— wow, the sunsets there were amazing—and what about that white rhino we tracked in the bush from sunset to sundown.* And all of this from wanting to remember to call Barry. It was then that I realized why my thoughts take so long to manifest, and the results of my creations are so circuitous.

—MARK

Associative Memory

With more possible neural connections than atoms in the universe, the brain has a big problem: how to find a memory. And if the proverbial tiger in the jungle is headed your way, or

Aunt Rosie, looking very tipsy, is charging your way, how does the brain find the right memory *quickly*? Emotions help you.

So emotions, which are themselves, in part, a neuronet, are tied into all the other neuronets. These connections allow the brain to find the most important memories first. They also insure that something important, like not putting your hand on the stove, is not quickly forgotten. It's why everyone remembers where they were and what they were doing when they heard about 9/11 and the World Trade Centers coming down, or President Kennedy being shot.

The up and coming "Emotions" chapter talks about how associative memory affects our behavior and reaction to the world, but there is one important brain-related function to go over. We said that emotions are, in part, neuronets. The other part is that emotional neuronets are connected to a small organ in the brain—the hypothalamus. This hypothalamus takes proteins and synthesizes them into neuropeptides, or neurohormones. And we all know what hormones do—at least everyone past puberty does. They prepare the body for action!

If it is a tiger, a hungry tiger, the hypothalamus secretes chemicals to get the body ready to run. Blood leaves the brain and the central part of the body and moves to the extremities— "fight or flight."

Emotions evaluate the situation quickly, in fact, without you even thinking about it, and send the chemical messengers off to fight or flight, smile or frown.

The downside with associative memory is that because we perceive reality, and treat new experiences based on our stored mental/neuronal database of the past, it's difficult to see what is really out there in the moment. Instead, the tendency is to just reference experiences from the past. This would seem to create a perpetual *Groundhog Day,* where the same old, same old happens day after day.

And who would the same old be happening to? Who would be reacting to situations based on the past? That vastly integrated cluster of neuronets that we've been calling "the personality." Just like all the cells of the body come together and interrelate with each other to produce a functioning organism, so the neuronets all interrelate, or associate, to produce that entity that we think of as our personality. All the emotions, memories, concepts and attitudes are encoded neurologically and interconnect, the result being what has been variously called the ego, the son of man, the lower self, the human, the personality.

In cases of split personalities, there are multiple integrated clusters, which are by and large not connected to each other. That is why when the personality shifts, there is no memory of the "other person." The cluster of nets that the person is operating from is not connected to those memories.

From this it becomes apparent why a hardwired brain results in a hardwired, unchanging, rigid personality. And while the personality may change from liking cappuccinos to lattes, that change doesn't really shift the personality to a new one. There's a million other nets that stay the same, thus the aggregate remains "you." Even though this sounds pretty grim, luckily the brain was created to take an incarnating spirit all the way to enlightenment, which is why it came equipped with neuroplasticity.

Neuroplasticity

Just as the Bill Murray character in *Groundhog Day* finally changes the behavior that is keeping him stuck in time, everyone has that option. It is possible to break the brain's wired-together neuronets, change habits and gain freedom. The key lies in the brain's natural ability to form new connections. Neuroplasticity is the term for the brain's ability to make those new connections—in other words, for neurons to connect to other neurons.

Where it was once believed that by adolescence the brain was pretty well wired for life, more recent research has confirmed that the brain is not only very plastic and malleable,

The brain likes surprises. After a surprise, the neuroplasticity in the brain goes way up. It's easy to see why: Suppose you're walking in the deep jungle, and Aunt Rosie in a leotard jumps out in front of you. Surprise! Your brain has to immediately go into high gear to work out a way to deal with a new situation. Connections have to fire instantly to link up all possible solutions and help you choose among them. You have to process the information very quickly in order to survive. Neuroplasticity also increases after laughter. And since neuroplasticity is the prime ingredient to learning, you learn better after a good laugh.

even into old age, but that it also *creates new cells.* As Dr. Daniel Monti explains:

> The good news is that there's enormous potential to change the kind of behaviors and characteristic patterns that we've fallen into. And the potential for change, within our nervous system, within our entire physiology, is tremendous.
>
> In fact, if you've listened and remembered anything that I've said, your physiology is different than it was before! That memory has been encoded, and your genetic structure has changed. And while we previously would talk about the nervous system as this very rigid thing that didn't have much capacity for change, we now know that on many levels, that isn't true. There's a tremendous amount of plasticity, which basically means ability to change, within the nervous system.

The research cited by Dr. Monti fits right in with the Human Potential Movement, which always said we are much more unlimited than anyone's ever realized. To think that our growth stops at adolescence is, in the words of John Hagelin, "a barbaric view of human potential."

> The Vedic tradition not only talks about the unified field, but very precisely describes it and provides experiential techniques, meditative techniques to experience and live it. And the practical benefits of living the unity of life are immense. There are hundreds of studies I could cite on the profound health benefits, the profound mental benefits, when orderliness of brain functioning is systematically developed.
>
> Coherent brain functioning results when we experience the unity within, and this coherent functioning of the brain translates to rising IQ, increased creativity, better learning ability and academic performance, moral reasoning, psychological stability, emotional maturity, quicker reaction

In doing brain research for the movie, we stumbled onto the bit about surprises and laughter increasing learning. Ah-ha! That's why we went into the scene with Marlee and the mirror (where the heroine, Amanda, experiences a self-hate meltdown then revelation) right after the humorous wedding party. The party was intended to give the audience a good laugh and a break from all the serious intellectual information. Swimming in neuroplasticity, the audience then rewires the brain to accept all that information, clearing the slates, so that it's easier to accept Marlee's experience as their own.

—WILL

time, greater alertness. Everything good about the brain depends on its orderly functioning.

And now orderliness of brain function can be longitudinally, systematically developed in students of all ages, even after the age of sixteen, when IQ traditionally is supposed to start to erode. "It's all downhill from there," it used to be believed. But that's not the case. It's a barbaric view of human potential. *We are meant to, designed to, engineered to evolve in creativity and intelligence throughout life*—but to do that you have to access the innate capability of the brain, and the tools, the key, to really develop the brain holistically is to experience the holistic reality, the meditative state, so-called spiritual experience, the experience of the unified field at the source of thought.

Frontal Lobe and Free Choice

The major factor that distinguishes human beings from all other species is our large frontal lobe, and the ratio of that frontal lobe to the rest of the brain. The frontal lobe is an area of the brain that enables us to focus attention and to concentrate. It's central to decision-making and to holding a firm intention. It enables us to draw information from our environment and our storehouse of memories, process it, and make decisions or choices different from the decisions and choices we've made in the past.

But many choices are far from free. Much of our behavior consists of conditioned, learned or automatic responses to stimuli. Dr. Joe Dispenza offers this example: "If you were in a dark alley and I threatened you, the normal choice you would make would be based on a physiological response of fear, which means your body mechanism would give you signals to run for your life or stay and fight." A similar process occurs when other neuronets kick in and produce automatic

Of course we have free will. Free will resides in our frontal cortex (lobe), and we can train ourselves to make more intelligent choices and to be conscious of the choices we're making.

—Candace Pert, Ph.D.

responses, like reacting to someone you know, lighting a cigarette or heading to the refrigerator when we are feeling stressed. These habitual, automatic responses hardly qualify as "choices."

A second way of making choices occurs when we consciously separate ourselves from the environment and its stimuli, and stand back from our habitual or biological behavior and become the observer. From this quiet vantage point, as Dr. Dispenza puts it, we can then "reason carefully, based on what we know . . . The frontal lobe takes information that we've developed through our lifetime, through experience and through factual intellectual data, and it says, I understand this neuronet, and I understand this neuronet, but what if I take those two neuronets, and I integrate those two concepts to build a new model, a new ideal, a new design?"

We're back to the observer. As Dr. Wolf notes:

> It seems mind-boggling that an observer would have any power in the world at all. In a certain sense, the observer has no power. In another sense, the observer has a tremendous amount of power. In the sense of no power, we would say, observations that are carried out, in a way that they were previously carried out, over and over again, in a repetitive sense. So, it gets to a point where we don't see the role of our observations anymore, because they become habitual. It's kind of like being an addict to something. You lose the power of observation. When you regain the power of observation, you can see that by your choices, you can actually alter, restrain, or change what you see "out there."

In the first scenario, the biological neuronets make the choice. The brain reacts to its environment, and certain aspects of the brain turn on automatic centers that cause the body to respond, like the blinking of the eye when an object gets too close or the classic "knee-jerk response" when the doctor taps your knee. In the second scenario, says Dr. Dispenza, "consciousness is moving through the brain, and using the brain to examine its options and possibilities." Instead of the brain going on autopilot and running us, we begin to use the brain. Consciousness begins to have dominion over the body.

Consciousness, the Observer, Intent and Free Will

We've examined these concepts before. And here we are again, inside the most complex structure in the known universe, the brain, once more looking at these concepts.

Remember in the quantum world how we: Intend to ask a question of reality (process 1), the possibilities (process 2) then arise, whereupon observation collapses them into a definite choice (process 3)? What Dr. Dispenza is saying is that it is possible to collapse choice into a new life: "Maybe we're just poor observers. Maybe we haven't mastered the skill of observation, and maybe it is a skill. And maybe we're so addicted to the external world and so addicted to the stimulus and response in the external world that the brain is beginning to work out of response instead of out of creation. If we're given the proper knowledge and proper understanding and given the proper instruction, we should begin to see measurable feedback in our life."

How do we translate this into action for change and transformation? From his scientific investigation into intention imprinting electron devices (IIEDs), Dr. Tiller has learned: "It is terribly important to sustain the thought and the intention if you want to make a transformation occur . . . when one wants to focus intent, you want to be a singleness of mind." And the reflection of that in the brain is the frontal lobe.

The tools for change and transformation are building . . .

And how you FEEL about that will ultimately determine which ones you use, and which ones stay in the tool chest. But that's the next chapter.

It does seem logical that the most malleable, complex, sophisticated physical device would be the interface between the "intangible" spirit world, and the "tangible" material world. And that it would mirror processes in both worlds. "As above so below, as within so without."

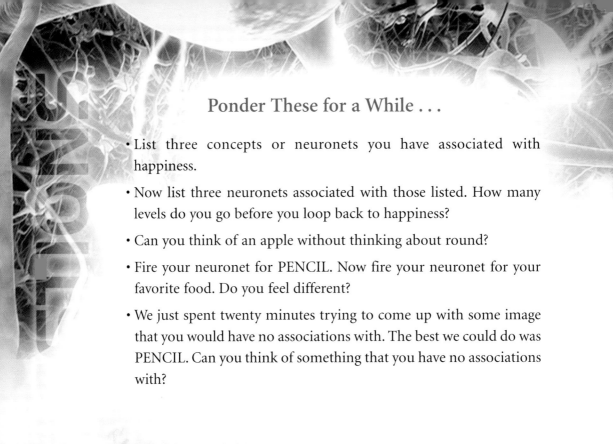

Ponder These for a While . . .

- List three concepts or neuronets you have associated with happiness.

- Now list three neuronets associated with those listed. How many levels do you go before you loop back to happiness?

- Can you think of an apple without thinking about round?

- Fire your neuronet for PENCIL. Now fire your neuronet for your favorite food. Do you feel different?

- We just spent twenty minutes trying to come up with some image that you would have no associations with. The best we could do was PENCIL. Can you think of something that you have no associations with?

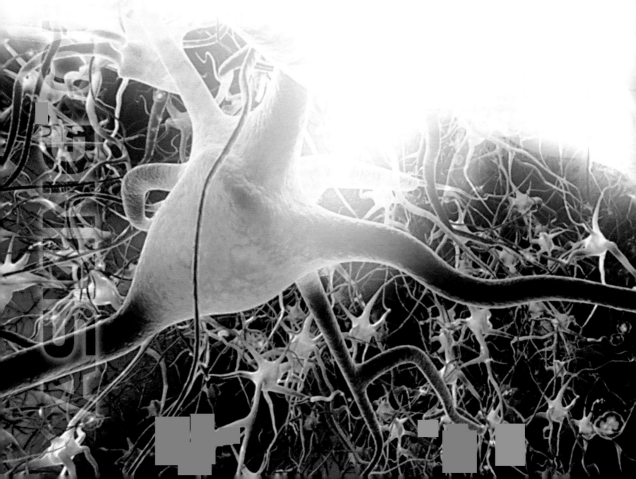

EMOTIONS

There is nothing either good or bad but thinking makes it so.

WILLIAM SHAKESPEARE

> # "Let's party!"
>
> ## —Pretty Much Everyone

Here we are, Emotions. Finally, we can have some fun! Enough of the egghead "Ponder This" brain-burners. We're out of brain class and into, well, fun! Emotions! Joy, sadness, hope, despair, passion, longing, winning, losing, on and on, the beat goes on.

Would there be rock and roll without emotions? Would there be you!? Let's think of all the stuff that would not be if there were no emotions:

- Beauty pageants
- Casinos
- Wars
- Poetry
- Victoria's secrets
- High school football games

In other words, we could go on and on to all the good, bad and ugly, wonderful, surprising and enriching aspects of human life. Would you ever laugh or smile if there were no emotions? Probably not. And you wouldn't even care!

Emotions—Mysticism or Biochemistry?

So what exactly are emotions? Are they some mystical property of experience that is indefinable, or are they something much more concrete and tangible?

One of the greatest parts of working with Will and Mark was being able to see myself and my emotions reflected back at me. Whenever I would get annoyed at Mark or Will for some attitude or behavior, I was astonished to realize that they were a reflection of my emotional state. When Will had his "Ah-ha" moment about creating situations where he could experience the "I told you so," I realized that I had the same issues. I've found if I can't put a finger on my emotional state—look around me, it's reflecting right back at me.

—BETSY

In the early '70s, Dr. Candace Pert fell off a horse. During recovery she was put on round-the-clock morphine. Being a scientist, she wondered how the drugs actually produced the effects that she was experiencing. So when the opportunity presented itself to investigate why they worked, she jumped on it.

It had been theoretically predicted that cells had "receptors" around the outside cell wall that chemicals "dock" to. The theory held that it was the chemical structure of the drug that allowed it to fit into those receptors, but no one had been able to find the actual receptors. Dr. Pert found the opiate receptors that line the cell wall. This discovery changed the face of biology.

"Once we'd actually found these receptors, then we started to wonder: Why would God put these receptors in brains if they were not there for some other purpose? Then after two seconds of thinking, a number of people around the world began to think that there had to be a naturally occurring substance made in our own brains. Well, about three years after the discovery of the opiate receptors, a team in Scotland discovered that the brain makes neuropeptides called *endorphins*."

Heard of endorphins? A.k.a., the runner's high? They are our own internally generated opiates. Further research ensued, and peptides started turning up everywhere. Says Dr. Pert: "In my laboratory at the NIH, I began mapping receptors for any other peptide that anybody had ever discovered in any biological system. And sure enough, whenever I looked for these receptors, we could find those receptors . . . We did a lot of detailed mapping of receptors and were able to see that not just opiate receptors, but these other peptides, were found in the parts of the brain that were thought to mediate emotion."

After this discovery, scientists began to see the receptors and peptides in a whole new light. As Dr. Pert says, "[We] started to think of these neuropeptides and their receptors as the *molecules of emotion*."

It became apparent that everything we feel, every emotion, produces a specific chemical, or collection of chemicals, that matches up with it. Those chemicals, or neuropeptides, or

Emotions are the chemistry to reinforce an experience neurologically. We remember things that are more heightened and more emotional, and that's the way it should be.

—Joe Dispenza

molecules of emotion (MOEs) are a chain of amino acids, made up of proteins and are manufactured in the hypothalamus. "The hypothalamus," Dr. Dispenza explains, "is like a little mini-factory, and it's a place that produces certain chemicals that match certain emotions that we experience." What this says is that every emotion has a chemical (MOE) associated with it, and it's the absorption of this chemical in our body by the cells that gives rise to the *feeling* of that emotion.

Pleasure/Pain

Not only did researchers find MOEs to match up with emotions, but they found them even in single-cell creatures. Dr. Pert found that what proved the molecules of emotion were, in fact, molecules of emotion was that, as she says, "We have the same identical molecules in the simplest one-cell creatures. And so, emotions are preserved throughout evolution. Endorphins are in yeast and tetrahina, simple one-cell organisms, so pleasure is so basic. And we were designed to run on pleasure. And I think we're addicted to pleasure, and our brain is set up to record pleasure and seek pleasure. And that's the final endpoint—to find pleasure and avoid pain. And that's what drives human evolution."

The hookup of these MOEs to what we perceive and experience is very direct. For example, the part of the brain that controls rapid eye movements and decides what to focus on is covered in opiate receptors. From an evolution perspective, this makes sense. We pay attention to what is important, and what is important, or what has the most meaning to us, is chemically conveyed, and quickly, to the body by these molecules of emotion.

Over time this simple pleasure/pain button has become overlaid with scores of other ideas, attitudes and memories. And even though it's a long way in evolution from an amoeba seeking food to French lace, emotions *had* to be wired into the body in a most compelling manner in order to solve the prover-bial "tiger in the jungle" scenario. And solve it quickly.

As an illustration of what goes on inside, and in keeping with our fun-chapter theme, the following "thought experiment" probes the workings of the memory/emotions/response interface.

ROBOTOMUS

Imagine for a moment that you were a little being who lived in a "biogenic robot"—Robotomus. You live in Robotomus' head in a little control room and look out through its eyes. Using complex levers, buttons and a computer, you feed Robotomus vital information.

And you have a job—to recognize what Robotomus sees, and interpret it so that Robotomus knows what to do. Now interpreting what that thing "out there" MEANS has nothing to do with moving parts; it's an abstraction—something from the realm of mind—something robots are incapable of computing. That's why you got the job.

Well, luckily, behind your little control chair, you have a massive wall of filing cabinets. They open and close based on what Robotomus sees "out there." So, you look out the windows of the eyes, and you see something! Suddenly, a whole bunch of filing cabinets pops out, and there's a whole bunch of folders that are glowing. All right! It looks like a bipedal humanoid. Next you check back through the eyes and see that it's a somewhat curva-ceous shape. Ah-ha! It's a woman! You turn back to the cabinets, and the folders pertaining to men close. Good, you've narrowed the choices down.

You then look a little closer to see what kind of woman . . . The woman is making a weird facial expression. Behind you all the filing cabinets but one snap back in.

So, in reality, we can't really say that we're seeing the world objectively as it is. There is no completely objective appraisal of anything, because our appraisal of everything has to do with our previous experiences and our emotions. Everything has an emotional weighting to it.

—Daniel Monti, M.D.

A single folder is glowing. You reach in and take it out. It says: "Aunt Rosie." You open it and look at the character history—something about her being abusive, cruel and violent.

You turn to the computer screen and the word: MEANING? stares back at you. A cursor flashes below it. Robotomus is frozen. You pound in . . . DEFEND, ENEMY AHEAD. Immediately, Robotomus starts shaking, and you look out the window and notice that the person "out there" is not Aunt Rosie, but displays a brief expression that is vaguely similar to Aunt Rosie's picture in the file. You reach frantically over to the computer and type: WRONG MEANING . . . MEANING UNKNOWN! But it's too late; there are chemicals being released everywhere, and it's getting extremely hot in the control room. Robotomus' legs are pumped with blood and adrenaline, but now it's shaking because it's gotten conflicting meanings and a ton of chemicals. You sigh, strap yourself in and decide you'll have to take Robotomus out for a run later . . .

———————

Sound familiar? First, there's recognition of a stimulus, then applying a meaning or interpretation to it, then instructing the hypothalamus to release neuropeptides into the bloodstream and boom! There's the feeling. What a beautiful system. So emotions are good, right? Absolutely. They're vital.

Excellent, so what's the problem?

As Dr. Dispenza explains, "We're making an analysis of every situation to determine if it's familiar, and that familiar feeling then becomes the means by which we predict a future event. Anything that has no feeling, we automatically discard or we reject because we can't relate to the feeling."

What's the Problem with Emotions?

The very beauty of the stimulus and response shortcut is the very thing that seems to trap us. Instead of evaluating a truly new experience from a fresh perspective, we tend to assume it's an experience we've already had.

When the same chemical events repeat themselves over and over again, the result is a cumulative emotional history. This history comes with identifiable patterns and predictable responses, which become embedded or "hardwired" in our brains.

That means our patterns and responses repeat without our having to think about them: stimulus-response-stimulus-response-stimulus-response. The survival shortcut mechanism becomes a trap into the same thing over and over.

Another "gotcha" is the hidden, buried and/or repressed emotions. Aunt Rosie may not always be mean; she just had a horrible toothache the day she snapped at you. Yet that neu-ronet is still in there and still gets fired, even though you are no longer aware of it.

Or forgetting it's the 21st century, the boss comes in, drops your report on the desk and comments: "Not a very good report." You panic, and the emotions run: Boss displeased → Loss of livelihood → Family unprotected → Barbarians invade village → Kill boss. And while it's doubtful you'll go after the boss with your mouse, your body has already responded to past situa-tions, and the chemicals are having their way.

One of the things I've been practicing recently is to move in and out of an emotion. In other words, if I can catch myself before I react to some-thing and head down the path of an unstoppable chemical cascade, I do this predetermined thing. I sort of halfheartedly move into the emotional feeling, and then I pull back almost like I'm outside of myself watching. I do this numerous times to practice being able to move between these two states. It helps train me to understand that I actually do have a choice. Standing back is like being a quiet observer. Moving in is like falling asleep into a dream I have no control over.

—MARK

And the Good News Is?

Survival for starters. Your emotions help you survive by giv-ing you a lightning–like reference that puts the puzzle together before you even know the pieces. And when you have a body, that's good news indeed. Going through life with emotions gives you a genuine experience of being alive, feeling, loving,

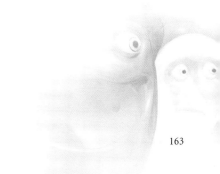

OWNING ONE EMOTION
One day I was on the phone
with Betsy, and we were going
over some animation stuff.
I was bitching about how some
of the animators were saying
what I wanted them to do was
crazy and that I was never going
to pull it off. I launched into a
tirade about how people always
tell me what I can't do, and that
it's happened over and over, ever
since junior high school.
I started marching back through
time listing all the "you can'ts,"
when I suddenly stopped.
Did someone say "repeated
emotion"? I realized I had
been creating this situation for
decades! Why? So I could then
have the "I told you so, I'm
smarter than you, kiss my ass"
emotion. I used it as a motivator
so that I could be better than
someone. And underneath that
was an insecurity! I was
projecting (unconsciously
creating) my own doubt. *Instead*
of all that drama and low-grade
motivation and satisfaction,
I could simply create.

This was the moment I finally
got what it means "to own an
emotion." And was it "retired"?
It's been three years, and no one
has told me what I cannot do.

—WILL

hating, living. Without those emotions, life would be boring. They're the spice in the (quantum) soup; the color in the sunset.

They provide much more than mere survival. They contribute to the ever-evolving evolution. That's evolution, not in the bodily sense, but in the non-physical, spiritual sense. Joe Dispenza says:

> Well, I don't have a scientific definition for soul, but what I will say is it is a register of all the experiences that we own emotionally. And the things that we don't own emotionally, we continuously reexperience in this reality, all other realities, in this life, all other lives. So, we don't get to evolve. If we keep reexperiencing the same emotion, and never retire that emotion into wisdom, we don't ever evolve as a soulful person. We're not inspired. You don't have the ambition or the desire to be anything else other than the product of the chemicals in our physical body that keep us on the wheel of living our genetic destiny.
>
> A soulful person overcomes the genetic destiny, overcomes the feedback from the body, overcomes the environment, overcomes their emotional propensity. Think about it. You want to evolve as a person, pick one limitation that you know about yourself and consciously act to alter your propensities. You'll gain something . . . wisdom.

In other words, those emotions may be pointing out something other than the tiger. They may be pointing out the pearl. Or rather they may be the grain of sand in the oyster that makes the pearl; that little bit of irritation that the oyster coats and coats and coats, until the pearl of wisdom is gained. It makes sense that the irritation, the pain, moves us to change. Happy, pleasurable emotions are not irritants. It's the *other* ones that either get repressed, suffered through or turned into wisdom— that fuller understanding of life and who we are. (Who am I?)

And beyond. Ramtha often asks his students when was the

last time they had an ecstasy, an orgasm, in the seventh seal.[2] Everyone is familiar with ecstasies in relation to sex (first seal), survival (second seal) and power (third seal), but what about those experiences in the higher centers? A profound new understanding, a revelatory "ah-ha", which also releases endorphins in the body, is an ecstasy of the sixth seal. The experience of cosmic consciousness, ultimate and intimate connection with God, is an orgasm of the seventh seal. Complete unconditional love is one of the aspects of the fourth seal.

According to this teaching, we never get to those dimensions because most of the time humanity is stuck in the first three seals: sex, survival and power. And the way out of the "bottom basement of humanity" is to take the emotions of the lower seals and own them into wisdom. Or as Dr. Dispenza puts it, "retire it into wisdom." Or as the oyster puts it, deal directly with the irritant until it is a pearl.

Certainly life without emotions is plain yogurt on bleached oatmeal for breakfast, lunch and dinner (with no honey). Repeating the same emotion over and over is blueberry yogurt and brown sugar on granola for breakfast, lunch and dinner. Our entire evolution is wired all our life with emotions—they are inescapable. So the question really is: How do we use them? What are we evolving them to? What are we becoming?

Passion, divine love, feeling oneness with everything, bliss and mystical experiences are all emotions—they generate those neuropeptides that flood the body and alter consciousness itself. A profound realization, one that doesn't have to do with the body—in other words, power, sex or survival—can so significantly rewire the brain that when people come back from it, they are different, and this world is never the same. Says Andrew Newberg:

---

[2] A chakra, or subtle energy center. They are nexuses of non-physical energy in the body. They line up with the ductless glands in the physical body and are said to be a key to unlocking higher dimensions.

One of the things that our research has been trying to show is that when people have a mystical experience, something really is going on in their brain.

It's not necessarily a delusional experience or hallucination, but it's something that is very real, in the sense that neurologically, something's happening. It's affecting us. It affects our bodies. It affects our minds, and how we ultimately respond to that, how we ultimately bring that information into our lives, that affects our behaviors and changes who we are as a person. Obviously, it has very real consequences for us as human beings.

A realization like this Ramtha describes as an abstract thought: "You are a God in the making . . . but some day you must love the abstract more than you love the condition of addiction. And if you love it first, the reality will manifest and your body will experience (it), and we have brand-new emotions, the likes of which you never knew before."

Brand-new emotions. All our emotions were at one time brand-new. And the reason we keep revisiting them is because they were so delicious. The pull of evolution is the possibility of a new set of brand-new emotions. Even more captivating, even more inspiring. Peeling off the layers of memory and habit to interact with a world that is now an exploding revelation. Now that's a party worth attending.

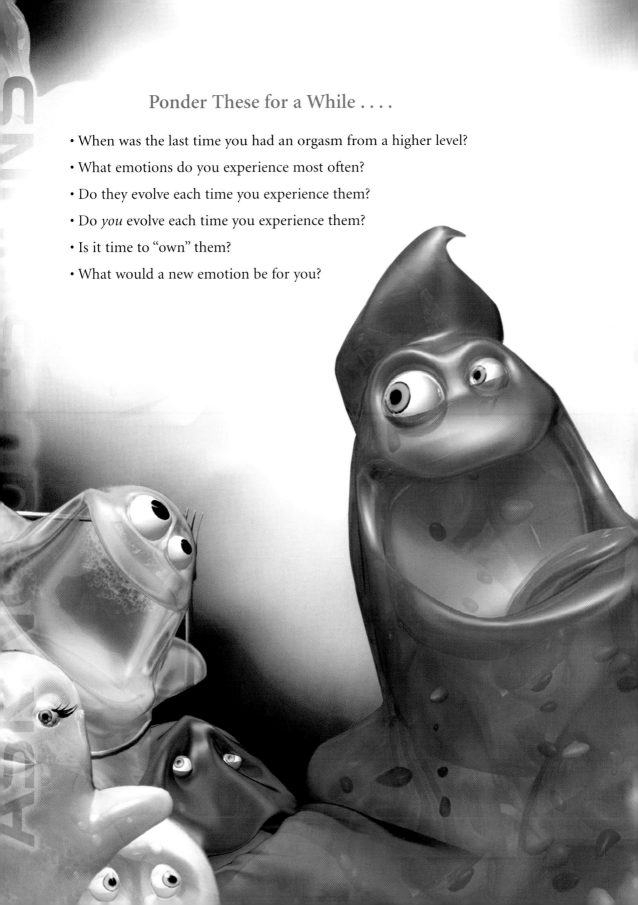

Ponder These for a While . . . .

• When was the last time you had an orgasm from a higher level?

• What emotions do you experience most often?

• Do they evolve each time you experience them?

• Do *you* evolve each time you experience them?

• Is it time to "own" them?

• What would a new emotion be for you?

This page is emotionless.

ADDICTIONS

The only difference between a rut and a grave is how deep it is.

CHARLES GARFIELD

Addiction is a suppression.
Do you know what that means?
It means that it suppresses you
back in your box.

—Ramtha

L et's take one of the most addictive drugs, heroin, to see how addictions work within the cells of the body. After it is injected, heroin docks with the opiate receptors of a cell. These are the same receptors that are biologically designed to receive endorphins, a neuropeptide manufactured by the hypothalamus. Instead of receiving the endorphins, the cell receives the heroin, and the cell becomes addicted to heroin.

Now let's try the same scenario with emotions. Emotions produce peptides, or molecules of emotion (MOEs), which dock in receptors on the cell. The same thing that happens with repeated use of heroin occurs with repeated use of the same emotion: Your body's opiate receptors begin to expect—even crave—that particular peptide. Your body becomes addicted to that emotion.

Shocking, huh? And "you like to think you're immune to the stuff." Drive by the alcoholic hanging on the curb, the junkies lined up in front of the meth clinics, the chain-smoker with yellow fingers and black lungs, and you might think, "Not me!" Think again; yes, you!

Shocking. But it explains *so* much! Do any of these sound familiar?

- Destructive emotional states
- Same situations over and over

- Inability to change
- Feeling helpless to create something new
- Deep cravings for certain emotional responses
- Voices in the head saying, "I want. Give me, give me."
- Saying you'll never do something ever again, then three hours later doing it

For all of the times you've experienced the above, this chapter is shock therapy. And it's for all of us (and that's all of us) who have neuropeptides coursing through our veins.

Humanoid (Noun): A Self-Aware, Self-Toking Organism

Through her research, Dr. Pert found that we have receptors that specifically receive marijuana. Why do we have these receptors? Because our body actually produces chemicals internally that give us the same type of high that we get from marijuana. This goes for whatever drug humans are physically capable of becoming addicted to—there is a chemical inside the body that is analogous to all drugs and a receptor for it to dock on. She explains, "We have marijuana receptors, and we have natural marijuana that's called endocannabinoids. Every time people toke up, their exogenous[1] marijuana is binding to receptors normally meant for fine-tuned internal regulation. So exogenous drugs plug into the same network meant for endogenous physiology self-regulation. These are the molecules of emotion. There's enough data now to suggest that no psychoactive drug works unless it's binding to a receptor that's normally meant for an internal juice."

In other words, any external drug that works in our body has an internal counterpart—that is why our body recognizes,

Heroin users have receptors for the heroin, and the more heroin that they take, their ability to make their own internal endorphins, their own internal heroin, basically starts to decline. Then their receptors start to become subsensitive where there's actually less of them, so there are these actual changes. Then there is this new information about how less brain cells are being made so that people get kind of, in all addictions, they get stuck in old patterns. They are just thinking the same thoughts over and over, and they are not able to think of something new.

—Candace Pert, Ph.D.

---

[1] *Exogenous chemicals* are external to the body. Endogenous chemicals are manufactured by the body and are "internal."

responds and becomes addicted to these drugs. External drugs use the internal receptors meant for the internal chemicals.

In the "Brain 101" chapter we covered how the emotions, and memories of emotional experiences, are encoded in neuronets and how these neuronets are connected to the hypothalamus. That is how we become self-toking organisms. All we have to do is fire up the correct net, and the chemicals start to flow internally. As Ramtha says:

> Addiction is the feeling of a chemical rush that has cascaded through the bodies through a whole assortment of glands and ductless glands. A feeling that some would call a sexual fantasy. It only takes one sexual fantasy for a man to have a hard-on. In other words, it only takes one thought here [the brain] for a man to have an erection in his member.

For many this is the most direct example of how focusing on a thought produces the proper neuropeptides. There are many other examples: remembering that glorious moment in high school when you caught the winning touchdown, the first time you realized you were in love, or success, like dreaming of the moment when the media calls you an inspired artist or wildly successful. In all these cases the frontal lobe holds the particular thought, activating the particular net that sends its signal to our internal pharmacy.

Does that mean that every time someone triggers this mechanism, they're an addict? Are you an alcoholic every time you take a drink? Of course not. If once a year you remember that glorious moment in the fall of 1972 when you got the touchdown, that's not an addiction. If every day you're wishing for those glory days—guess what? You've got a habit going.

Biological Effects

It's common knowledge that addictions have serious long-term effects on the body. With the discovery of the peptide-receptor mechanism, the biological basis of the effect of addiction, as Dr. Pert explains, has become obvious:

> If a given receptor for a given drug or internal juice is being bombarded for a long time at a high intensity, it will literally shrink up; there will be less of them or it will be desensitized or down regulated so that the same amount of drug or internal juice will illicit a much smaller response. The best example of this that people are familiar with would be tolerance. We all know how an opiate addict has to take larger and larger doses of drugs to achieve the same high.

The same tolerance effect is seen in emotions. The thrill seeker who pushes himself further and further into extreme bungee jumping from airplanes to get that adrenaline high, or the sexaholic who seeks the kinkier and kinkier, or the politician running for higher and higher offices, not out of a desire to serve, but on a quest for more power. If you begin to look for these scenarios with people you know, or especially, in your own life, you'll see examples everywhere.

Meanwhile those poor little cells are being starved. The constant overuse of the chemicals required to produce an emotion, like anger, in the body result in desensitized receptor sites being created to adapt to all those anger neuropeptides. The cells are no longer getting a "well-balanced" meal, as they receive whatever emotion they are addicted to more than others, so they are left with having to get a narrower supply of nutrition. The more anger the personality creates, the more satiated the cell will feel. This is the story behind a guy who on Friday night is out "looking for a fight." He's not angry for a particular reason; he's just out feeding his little cell friends. And those little guys can raise quite a racket when they need something. Ever hear a little voice in your head that says, "I'm hungry" or "I'm thirsty"?

Ever wonder who said that? Well, according to Ramtha, those voices in your head are the collective voice of your cells. They are telling you, *"Feed me."* And if the emotion is one that you think is socially or morally incorrect, you won't hear, "Let's go

Everyone is addicted. I don't care who they are. And they're addicted because they've never had anything better to replace what they are addicted to and have a reason to wake up every morning to live for.
The man who is addicted to power gets up every morning, and he drives things that show his power. He has to have a lot of people to feed off of, to suppress, to order around, in order to feel worthy. Because he does not feel worthy. He needs the emotions to feel worthy.

—Ramtha

When we fast from those emotions, those voices that come up, the cells are literally sending nerve impulses up to the brain, to let it know that it's starving, the body is starving, from what it's depended on chemically. And those chemicals are very powerful information carriers.

—Joe Dispenza

make someone feel stupid so we can feel intellectually superior," but rather there will be a nebulous craving that makes you unconsciously go out and make that happen. In other words, you'll create it.

Emotional addictions explain so much—why someone constantly trashes other people, or gets into the same abusive relationship, or the same horrible living situation over and over. In other words, emotional addictions are why people continue to create a particular reality in their lives, even though they say, "Well, I would never create *that.*" The only way to move past these repetitive behaviors and addictions is to say, "Well, I do create *that* over and over, so I must be addicted to it."

For many, many people, all the creations in their lives are emotionally, addictively based. As an example of creating something "bad" in your life, let's look at "victim mentality." Something bad initially happened, you told someone, *they felt bad for you* (now they're suffering, too) and so they fixed the problem. Relief. *Hey, not bad,* you might be thinking. *Let's see if I can make that work again.*

Suddenly, people are taking care of you. They're giving you money, supporting you emotionally, are sympathetic and available anytime you need them. Of course, the downside is that there is a built-in shelf life to the victim/savior relationship. Every savior needs to feel just as special as the victim, so they tend to move on after the initial "rush" wears off. If either of you don't change, both of you just go on to rediscover your addiction with someone else and someone else. And keep on, in the words of a true victim, "having bad, unfair situations happen to me." Sounds rather different from, "I continually create situations that allow me to get sympathy and support from people."

Dr. Joe Dispenza put it eloquently when he said: "My definition of an addiction is something really simple: It's something that you can't stop. If you can't control your emotional state," he says, "you must be addicted to it."

Emotionalics Anonymous

In some ways it's a pretty grim picture. I'm addicted; you're addicted—let's get together and rub our addictions together. Actually, that doesn't sound so bad; that's what everyone does all the time. We're frequency specific with those emotions, and so bring like-minded beings into our sphere. According to Ramtha: "The people that we really love are people who are willing to share our emotional needs, our feelings." Dr. Dispenza describes it this way: "We're breaking the addictions to all those agreements chemically. That is an uncomfortable state for the human . . . because you look for some evidence in your life that you are doing the right thing, and everywhere you look for evidence in your life is with the people you've had all those agreements with."

But still, it's grim because addictions are hard to break. That's what makes them addictions. And the way these emotions got to be an addiction is through continually trying to recreate an initial experience. The first experience of sex, or sympathy, or power is not an addiction. It is chasing that high over and over that becomes the addiction. Says Ramtha:

> Now, what about people who are addicted to sex, to heroin, to marijuana? Well, they all do different chemistries in the brain. They're endeavoring to touch the pleasure center in the brain. *That is not what that brain was meant to do.* So people reinvent experiences in their brain by distributing the same chemicals, the same feeling.

And what was the brain meant to do? To dream new dreams, new realities, then bring them forward into manifestation and experience that first incredible emotional moment . . . a moment with a new emotion.

Sounds great—new emotions, new highs—so why is it so hard to break the habit?

HOOKED ON A FEELING

Science knows now that the hypothalamus makes neuropeptides, and those neuropeptides are strong chemicals. For example, they have taken laboratory animals, and they've hooked up electrodes in certain parts of their brain that produce those neuropeptides. They then train the laboratory animal to press a lever to get that chemical release, that neuropeptide release.

It would choose the neuropeptide release more than hunger, more than sex, more than thirst, more than sleep. As a matter of fact, it went to the point of physical exhaustion and collapsed, before it would take care of itself physically. And that's really what stress does to our body. We become so addicted to the stress in our lives that we can't quit our job, even though it doesn't serve us. We can't leave our relationship, because it doesn't serve us. We can't make choices, because the stimulus and response are producing the chemistry that clouds our choices. And we're no different than the dog, who lacks its ability to make choices because of its smaller frontal lobe.

—Joe Dispenza

You can't cure an addict until you give an addict everything they want and then they ask no more. That's when we have owned an experience. And that is when we are wise. Couple that with indeed new revenues to the mind. And revenue to the mind is knowledge. From knowledge, it's like building blocks; we build new holograms, and we create realities.

—Ramtha

And we're still asking the question of how to break those addictions!

Possibly the most successful program ever in dealing with addictions is Alcoholics Anonymous (AA). Millions have kicked their alcohol addiction "one day at a time" through implementing their 12-step program. It would be a disservice to attempt to go over it here, and anyone interested should check it out.

But take a moment and consider this part of the program—the alcoholic is told to repeatedly reaffirm, "I am an alcoholic." And while in the beginning that is necessary to face up to the reality of the situation, it forever locks the person into that personality. So it's never over. The addiction has never been owned and retired. The person continually identifies with that which he is trying to push away and release. Finally, it denies the person the possibility of complete and utter transformation, which is why we're here.

Why ARE We Here?

Ahh, back to the Great Questions. And why are they great? Is it because they are not obvious or easy to answer? Or because they seem meaningful? Or because they sound great at a cocktail party, and you get to impress people when you bring up Great Questions? It's because they're the answer to get one out of a great mess.

> We are here to be creators. We are here to
> infiltrate space with ideas and mansions of thought.
> We are here to make something of this life.
>
> —Ramtha

> Our purpose here is to develop our gifts of intentionality.
> And learn how to be effective creators.
>
> —William Tiller, Ph.D.

> The point is that we're here to do something with ourselves.
> We're here to explore the total limits of creation;
> we're here to make known the unknown.
>
> —Miceal Ledwith

> The whole purpose of this game: We prepare our
> body chemically, through a thought, to have an experience.
> However, if we keep preparing our body chemically to
> have the same thoughts, to have the same experiences,
> we don't ever evolve as human beings.
>
> —Joe Dispenza

Emotions that we've depended on for so long are now no longer being given to the cell, and the cell goes in decay on us. If we persist past it, just like we persist past any addiction, we break the response, because we're not responding to the voice in our head. At the same time, we're breaking the response chemically, because now the cell isn't getting its chemical needs. The cell, ultimately, will be released of its chemical addiction, and now when it reproduces itself, it "up regulates." It lets go of all those receptor sites that have been responsible for those emotional states, and now the cell is in a better state of harmony, and the body experiences joy.

—Joe Dispenza

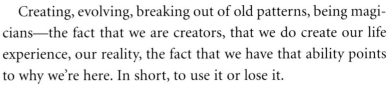

Sixty to eighty percent of crime is linked to drugs and addiction. Just think of the possibilities for change, not just on a personal level, but on a societal level as well.

Creating, evolving, breaking out of old patterns, being magicians—the fact that we are creators, that we do create our life experience, our reality, the fact that we have that ability points to why we're here. In short, to use it or lose it.

If we are here, as Dr. Ledwith says, "to make known the unknown," that would mean experiencing for ourselves something we've never tasted before. The same old, same old becomes the ever new, ever new.

Addictions are broken by changing, evolving.

And if that is why we are here, those new emotions will be so amazing, fulfilling and luscious that the old ones will seem like an old high-school yearbook—a big deal at the time, but now retired to a forgotten bookshelf. Dr. Pert reports and biology supports this transformation through very recent discoveries. There is evidence now that when people or laboratory animals, like rats, are addicted to a drug (nicotine, alcohol, cocaine, heroin), all of the test subjects have something in common—growth of new brain cells are blocked. But when the subjects discontinue taking the drug, the new brain cells continue growing. As Dr. Pert says, "One can completely recover, and one can make up her mind and create a new vision for herself, a new brain." There is hope in a new beginning for many, from the smallest addictions, to the largest.

As Ramtha sums up the way out: "We must pursue knowledge without any interference of our addictions. And if we can do that, we will manifest knowledge in reality, and our bodies will experience in new ways, in new chemistry, in new holograms, new elsewheres of thought, *beyond our wildest dreams.*"

Ponder These for a While . . . .

- Why does it feel sooo good to feel sooo bad?

- List some of your emotional addictions.

- Okay, so what's the addiction you didn't list?

- List the addictions of the people closest to you.

- How are you able to recognize their addictions?

- Are all addictions bad?

And that's the final endpoint,
to find pleasure and avoid pain.
That's what drives human evolution.

—Candace Pert, Ph.D.

DESIRE → CHOICE → INTENT → CHANGE

It really is a question of desire, desire, desire, and only desire.

It has nothing to do with ability, talent, brains, or anything else.

FRED ALAN WOLF

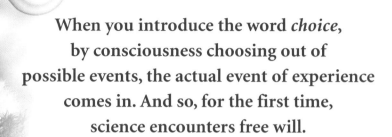

When you introduce the word *choice*,
by consciousness choosing out of
possible events, the actual event of experience
comes in. And so, for the first time,
science encounters free will.

—Amit Goswami

So, in effect then, when one wants
to focus *intent,* you want to be
a singleness of mind.

—William Tiller

The good news is that there's enormous
potential to *change*.

—Daniel Monti

D esire has a bad rep in spiritual circles. Phrases like "kill out desire" make it seem that if a being is without desires, then enlightenment will quickly follow. The second of Buddha's four noble truths is:

The origin of suffering is attachment to transient things and the ignorance thereof. Transient things do not only include the physical objects that surround us, but also ideas, and in a greater sense, all objects of our perception. Ignorance is the lack of understanding of how our mind is attached to impermanent things. The reasons for suffering are desire, passion, ardor, pursuit of wealth and prestige, striving for fame and popularity, or in short: *craving* and *clinging*.

A quick read of this seems to place desire, passion, ardor, etc. on the list of culprits. Therefore seekers strive to remove those emotions from their minds and live a desireless, passionless life. However, *craving* and *clinging* give a clue to the real culprit— it's not desire, but the attachment to desire. (Note also that it's not bad, just ignorant, or "the lack of understanding of how our mind is attached.")

"Attachment to desire" sounds a lot like "addiction to emotion." In fact, it is the same thing. Try interchanging the two expressions to verify that for yourself.

This is one of those instances where science has crossed paths with the great spiritual teachings and rediscovered the same phenomena. According to Buddhist traditions, it is this attachment that keeps us forever on the wheel of life, death and rebirth, going round and round, over and over. As Dr. Pert put it, "In all addictions you get stuck in old patterns. You are just thinking the same thoughts over and over, and you are not able to think of something new."

Desire and Passion—Friend or Foe?

Desire and passion fuel evolution and change. Dr. Dispenza says, "You have to have the willingness and the passion to step outside the boundaries of your own comfort zone." Certainly this scene of Jesus is one of passion: "And when he had made a scourge of small cords, he drove them all out of the temple, and the sheep, and the oxen; and poured out the changers' money, and overthrew the tables" (John 2:15).

In analyzing a given desire, it's important to do two things: don't judge and be honest about what the desire is really about. To look clearly at a desire, judgment has to go, for it's the labeling of a desire as bad that sends it to the basement of

Desire is a mechanism for me to examine myself and reflect my understanding into reality for experience. Through whatever desire I am having, I can examine why I have that desire. I ask questions like: What will achieving that desire fulfill in me? Is it because of an emotional addiction? Will achieving this desire bring me a new experience? Will I be playing out an old experience? When I can examine my desires for agreement this way, I can see why I have them, and it helps me become clear in my intent. If my desire is for power, and I acknowledge that I want to experience power from a new perspective and why I want to experience power, then I can have that experience and move on from it, because I know truthfully what I want and why I want it. Clarity and honesty bring the experience in a way in which I know I can gain knowledge from it.

—BETSY

repression. Desires arise. Someone cuts you off on the highway—for a moment the desire is to have a side-mounted laser cannon and blow him off the road. If you feel horrible and ashamed, there's a very good chance that the cause of that anger will never be discovered.

As for what the desire is really about, take an example: someone running for a public office. They desire power. But often people feel guilty (judgmental) about so boldly stating their true desire that they go into a song and dance about helping the citizens, whereas they really want to experience the feeling of power. And who's to say that for them that's not the next step up their ladder of evolution? Or maybe they say they want power to cover up a deep feeling of insecurity and worthlessness. In which case getting the power won't really do them any good.

There's also a very practical reason why it's imperative to get at the root of desire. Manifestation! As Bill Tiller mentioned, you want a singleness of mind. If the true desire is being overlaid with a politically correct one, or a desire has another desire underneath it, that means that two neuronets are being activated. It's a house divided, and that just doesn't work with intent. But before that, of the unending (or so it seems) desires, which bubble up from all the corners of the self, the big question is: Which one should we hit the "go" button on?

Choice

Someone has to choose. And who chooses? For simplicity, let's say it's one of two entities. One is the personality, and the other is the transcendental self. This way of organizing your two selves brings us back to the ego/god, matter/spirit split. We know that if it's the personality, the choice is coming from the preexisting neuronets, meaning past experiences and emotions, and the addictions thereof. In which case the "go" button can be relabeled the "repeat" button. And more often than not, this choice comes from an unconscious decision, just like the laboratory animals that keep pushing the peptide lever.

The true "go" button comes from the spiritual side. The choice in this case is not motivated by the past, but making known the unknown, or evolving. But given that, an interesting question is—of the desires that arise, how do I know which are ego-based and which are soul-based? Especially when one considers that the ones that arise from the spiritual side are often a little strange, outside-the-box, or "crazy and weird" when compared to our normal routines.

There are some great examples of this in the stories of students and spiritual teachers. In many of these relationships, the teacher is the voice of the student's sleeping spiritual nature. The idea being that the student on his own would take millennia to hear his own inner voice, so the teacher communicates it for him. Typically we think war is bad, right? But in the example of Krishna, he is driving Arjuna's war chariot, telling him it's his spiritual task to go and war with the Kurus.

Another example is the Buddhist story about Marpa and Milarepa, who build a huge stone house. Upon completion, Marpa had Milarepa put every stone back where he got it, deconstructing the entire house. This might sound crazy, but not as crazy as the fact that they repeated this dance four more times.

Don Juan had his student Carlos Castaneda, who was by now a best-selling author, flipping hamburgers in a diner for months. Did Don Juan intend enlightenment through burgers and fries? It turned out after a few months, a beautiful young woman came in looking for Carlos. Carlos kept quiet until a big limo pulled up. The woman said, "It's Carlos," whereupon Carlos realized how he was still craving fame, and in that moment he figured out his true colors.

It's often that voice coming from the transcendental side, that crazy desire for wisdom, that makes a transformation no one would have been able to guess. This is why it is so important not to judge desires, but to look closely at the desires

THE ENTERTAINMENT INDUSTRIAL COMPLEX

In the '60s we all talked about the Military Industrial Complex (MIC). And while that is still an operative force, quietly and steadily over the years it has been supplanted by the Entertainment Industrial Complex (EIC). This impacts everyone's life moment by moment a hundred times more than the MIC. You see, the EIC are masters of two important things: manipulating desire and choice and disempowering people. They use "entertainment" to create the desire and the emptiness that make you buy what they produce. And although in principle it's not that different from the Roman coliseum, with the reach of technology, it's overpowering. Just observe how every news show, magazine, movie, TV show creates desires and/or makes you feel powerless and empty—feelings that only products can fill.

—WILL

> The best way that a human being can become that observer is, first of all, having the awareness or the understanding intellectually that they don't always have to make the same choices over and over again. And second, be put in certain situations, experimentally, in your own personal life, where you actually override those mechanisms in your body, and that takes practice.
>
> —Joe Dispenza

before choosing. And then there is the choosing.

"Free will resides in our frontal cortex, and we can train ourselves to make more intelligent choices and to be conscious of the choices we're making. And I think it takes practice, different kinds of practice. We can go to the gym and pump up our bicep, or we can pump up our frontal cortex by doing yoga and meditation and other practices," Dr. Pert tells us.

So *who* chooses? Of course, *you* do, but then it's once more back to the question, "Who Am I?" Which you (personality/ego or transcendental/spirit) does the choosing? Neurologically, it seems the question is: Is the choice coming from an existing neuronet, or is it coming from the frontal cortex? Are we taking advantage of that quantum randomness bubbling up through the Chinese boxes[2] to let us choose something new, or are we the mechanistic machine that performs everything based on preexisting (old) conditions?

And once more back to the question: Which world do you live in? The living, organic, alive, interconnected universe, or the windup toy soldier, ticktock, divided one?

It is your choice.

Intent

And on the other side of choice is action! When Dr. Tiller did his research on intention affecting physical systems, he used "four very well-qualified meditators, highly inner-self-managed individuals." As Dr. Pert suggests, this ability to wield intent is a skill that can be developed. Dr. Tiller adds in regard to the ability to focus intention: "That's why some of the old occult teachings teach people to focus on a flame. You learn to bring your attention into a very sharp channel, so that the energy density becomes greater."

At this point, we're *still* talking about creating reality. But

---

[2] From Jeffrey Satinover's quantum brain theory on free will.

DESIRE ➔ CHOICE ➔ INTENT ➔ CHANGE

now it's about why we're creating what we are, what levels we're creating from and how to make those creations more conscious and more powerful. It seems our amazing brain is wired in the frontal lobe for exactly that purpose. Says Dr. Dispenza:

> What separates us from all other species is the ratio of our frontal lobe to the rest of the brain. The frontal lobe is an area of the brain that's responsible for firm intention, for decision making, for regulating behavior, for inspiration. And as we develop that ability as human beings, we make other choices that, in fact, will affect our potential, or affect our evolution.
>
> What may take a canine, or a dog, literally thousands of years to make another choice, the human being, because of its enlarged frontal lobe, can make the choice within a matter of moments.

So we've talked about why desire is necessary, and not always "bad," and that's why you must make the executive decision and line it up with intent to bring into reality health, wealth and happiness. If you're still feeling a disconnect, Ledwith puts forth, "Now, why can I not achieve those things? Fundamentally, through lack of focus. We cannot stay focused; the mind is wandering all over the place, for very long, and we're too attuned to the vibrations of this material plane." There's the rub. For intent to really work, it has to be focused—but the world in which we live is always, always grabbing for your attention. Billions of dollars get spent to get you to pay attention to what they want you to focus on. (And the operative word is *pay*.) It's a real dilemma, notes Dr. Dispenza, as "most people stop because they look for the results in the world after a small amount of effort, and when they don't see those results, they immediately discount its possibility. And yet, right on the other side of that, right past that point where they stop is the potential still existing. We're lazy as human beings. We live in a convenience world, and if we don't get our quick, immediate fix

We are running the holodeck. It has such flexibility, but anything you can imagine it will create for you. Your intention causes this thing to materialize once you're conscious enough and you learn how to use your intentionality. You learn to control intentions. And it's all this building process, more and more and more, so that you then experience these deeper signals from these various layers of the bio bodysuit, and you sense the larger reality.

—William Tiller, Ph.D.

The only thing constant is change.

—I Ching

If reality is my possibility,
the possibility of consciousness
itself, then immediately comes
the question of how can I
change it? How can I make it
better? How can I make it happier?
In the old thinking, I cannot
change anything, because I don't
have any role at all, in reality.
Reality is already there.
It's material objects moving
in their own way, from
deterministic lives.
I, the experiencer, have no role
at all. In the new view, yes,
mathematics can give us
something; it gives us the
possibilities that all this
movement can assume.
But it cannot give us the actual
experience that I'll be having
in my consciousness.
I choose that experience,
and therefore, literally, I
create my own reality.

—Amit Goswami, Ph.D.

on what we need, we're impatient."

But, of course, we can't really blame the world for our lack of focus. That's victim mentality. Rather, in order to get really good at exercising *intent,* we must *desire* to get good at it and make the *choice* to develop it. It's a chain reaction with extra-ordinary results.

In the "Quantum Brain" chapter, we talked about the Quantum Zeno effect—when one is continuously holding (i.e., focusing the mind) on the same intent directed toward the quantum world, reality is affected. To requote Henry Stapp: "By virtue of the quantum laws of motion, a strong intention, manifested by the high rapidity of the similar intentional acts, will tend to hold in place the associated template for action." So it's not just having an intent and going to the movies, but the repeated desire desire desire, focus focus focus that makes the magic happen.

Observable Change

We've been talking for chapters now about the glories of change, and the dead end (literally) of never venturing into the unknown. So by now either you're itching for some mind-blowing experience, or it's time for bed.

To wrap up this long chain of operable and interrelated factors, we asked Dr. Dispenza to bottom-line for us how to make it all happen for everyone. As he told us in an earlier chapter, "The subatomic world responds to our observation, but the average person loses their attention span every 6–10 seconds." How do we overcome that?

QUANTUM: FROM THE TINY
TO THE HUMAN

So how can the very large respond to someone who doesn't have the ability to even focus and concentrate? Maybe we're just poor observers. Maybe we haven't mastered the skill of observation, and maybe it is a skill. And maybe we're so addicted to the external world and so addicted to the stimulus and response in the external world that the brain is beginning to work out of response instead of out of creation.

If we're given the proper knowledge and proper understanding and given the proper instruction, we should begin to see measurable feedback in our lives. If you make the effort to sit down and design a new life and you make it the most important thing, and you spend time every day feeding it like a gardener feeds a seed, you will produce fruit. It may take some time in the beginning; it may take some time to develop the art to control your mind from all that it needs to do.

But we should be willing to sit down every day and take a piece of our day and set it aside and give thanks that we're alive and give thanks for the life that we have. And then begin to observe by pure sincerity to design a new possible future for ourselves. If we do it properly and we observe it properly, we should have opportunities begin to show up in our lives—not equal to what we can predict, but outside of our prediction. It must be outside of our prediction because when it is, we know that it has come from a greater mind.

If it's equal to our prediction, then we are creating more of the same. How can we create a new world if we can predict the outcome? We're not creating anything new [when we] look for feedback in our world. The brain loves feedback. It likes to know when it has succeeded or has actually accomplished a task so that it'll do it again.

You can't just think yourself into change. You can't just use the same brain that got you in the pickle to get you out of it and make a change in your life. You have to move outside yourself to gain a greater perspective. For some people, it can be as simple as taking a bath and contemplating things, or being in nature, or focusing, or meditating, or doing a fire walk . . . whatever it is for each person that puts them in a different state.

—MARK

That sincerity, in that level of mind, gives people the opportunity then to apply some principle of choosing a potential reality and observing it. And by doing it, they are willing to say, I know quantum physics works for the very tiny.

But when do you become a maverick and say [what] all of the greats that have ever lived have said? It always starts with the thought and the dream. And why not get about the business of applying it to your own personal life? It's not that the quantum field doesn't respond; it's that we have to up our level of will and sincerity; and when that happens, we'll see measurable results in our lives.

—JOE DISPENZA

So to see what is necessary for change, we simply have to walk back up the path we just walked down. To change, we intend the change. Intention is the result of a decision (free will) to change, and that decision arises out of a desire to change.

You gotta want to change. We mean *want* to change, desire it like you desired your first . . . whatever. For the material world, the world of matter, runs like a clock and resists change, while the unseen world of the spirit calls it forth. The choice is which world to live in.

I've seen some really dumb people turn out to be excellent scientists, even winning Nobel Prizes. I've seen some remarkably smart, talented people begging for money in the streets of San Francisco. It's desire, desire, desire.

—Fred Alan Wolf, Ph.D.

DESIRE → CHOICE → INTENT → CHANGE

Ponder These for a While . . .

- What is it that you desire and wish to be true?

- Why do you desire this?

- Where does this desire come from?

- What will achieving that desire fulfill in you?

- How would your reality change if you achieved this desire?

- Would your paradigm change, too?

- Are you willing to give up everything in your current paradigm to achieve this desire? Do you have to?

"What does all this really mean to me?"

While making *BLEEP*, I made everyone crazy asking this question over and over again. How do I integrate all this new information into my life to effect actual change? How do I take it from philosophy to experience? When I started this film, I didn't even know how to spell quantum physics or care much for "spirituality." I was living happily in my shoe consciousness, bouncing off reality. I think of myself as the "accidental creator."

In the last four years of working on the film, this book and traveling around the world speaking about it, I've come to understand it this way: Emotional addictions (or attachments) seem to connect back to the beginning of everything. I create my reality based on my emotional state, which is a state that I choose because my body has become addicted to the same experiences/ emotions/chemicals. Those experiences are wired into my brain based on old experiences, old data.

So how do I get from my emotionally addicted state to a higher state or higher level of myself and from there create new emotions? How do I "own" an emotion? What does that mean?

For me, "owning" an emotion means it no longer has power over me or my choices. I choose my emotional state; it doesn't choose me. Once I've "owned" an emotion, it doesn't mean that I'll never have that emotion again. When it comes up, I don't panic and try to repress it. I don't let it run havoc over my system. I observe.

"How do I begin—where do I start?" Well, if emotional addictions are where I have been creating my reality from, then I'll look at those. I try and work on just one, but usually I find that many are interconnected and can be traced to one or two experiences I had earlier in life.

So first I realize I have them—acknowledge them.

The next step is to get beyond the judgment. I found that once I discovered an addiction, I would spend time feeling bad about it, judging myself, and the funny thing is, I would judge myself from my addicted state. Here's an example: I'm addicted to failure. And I judge myself as not being good enough to get over failure (which is failure!). Ugh! I remind myself that I'm not alone in this problem. Everyone is addicted to something, and everyone, including me, has the ability to change their addictions. But judgment is one of those issues that is wired deep. Guilt and judgment seem to be wired into us from centuries of "morality" being pushed upon us. So even if you think you're not a judgmental person, it's there somewhere.

So what do I do?

I spend a couple of days writing down every emotion I feel, every time I feel it and the event that it is connected to. This is an eye-opening exercise. Not that I feel qualified to hand out exercises (oops, failure!), but this really helps me see my addictions. I do this every couple of months.

Once I had a list, I began to interrupt my pattern. Every time I began to feel an emotion that I was addicted to, I would stop and ask these questions:

- Do I really need to go forward with this?
- Who/what does it serve?
- Will this solve the problem?
- Why do I look at this as a problem?
- Will it evolve me?

Answering those questions allowed me the time in observation to begin to see how I can choose my emotional state—to affect my reality in a way that moves me forward. Something else I realized is that there are many layers to emotional addictions. Anger is a by-product of resentment, which is a by-product of failure, which is a by-product of victimization, and so on.

After getting over the "This is impossible, I'll never resolve this" emotions, I actually saw the fun in it. It was a new experience. Rebuilding myself from the inside out instead of feeling like the outside was building me!

It's not always the big things or the obvious. You have to look deep and at the very small to lead you to the big (sounds a bit like quantum). I look into my environment—my reality—my people, places, things, time and events. From observing these things, I can see how my addictions have created my reality. I can also see how these addictions continue to manipulate me into certain situations.

Example: At one point in my life, I was late for work every day. No matter how early I got up, or what I did to try and get to work on time, I was late. I always got stuck behind this school bus. Every day it was the same damn bus, which made me feel stressed out and rushed and overwhelmed at work. I felt like I never could get anything done. I was cranky, and everyone would walk on eggshells around me. I was in my midtwenties and had a job that most people would expect an older man to be doing. I had a complex. *I don't deserve this job; I can't do this; no one respects me. I mean, if I can't even show up on time, I shouldn't be here!* Now I see that the bus was my own creation to help me fulfill my chemical needs of failure! I had a good laugh over that one.

As I said, I do this every month or so to see my progress. To check in with myself. When I'm conscious of my thoughts, my emotions and I ask my questions, it's as if I can stop time—go out into the future and sample my possible choices, then make the choice that will best evolve me. Maybe it is my true desire to have that experience.

I also make time to sit every day and consciously create my reality. But it's not that I sit down and say, "Today I want a million dollars to fall from the sky." I focus on abundance and allow that my god will bring it to me in a way that will evolve me. Because that is what it's about—evolving to a higher level of consciousness. Which means that I have emotions and experiences I need to have and then own. By taking the time to observe myself, my reality, my emotional state, I can make choices that will move me along a greater path, instead of just bouncing off the walls and buying shoes.

We have received so many letters from people around the world who have taken this knowledge from philosophy to experience. There are some great stories to inspire you! Here are a few:

About four years ago, my husband passed. He left behind him a bird deep in grief, not to mention his wife. No matter what I tried, the bird continued to bite, scream, pluck his feathers and be very self-destructive. Finally, after watching the movie, I wondered if I changed my thoughts and actions, prayed for him and believed, it could be different. It started about two weeks ago. He lets me hold him, bathe him, play with him, and he does not bite. This is amazing [because] this bird has always bitten. He is happier. I am amazed that by changing my thoughts and actions, I impacted his. Before I was desperate because I didn't know what I was going to do with him. Now he has a brighter future.

—JEAN

I suffer from pretty bad hay fever. My dad and brother do as well. We all have it so bad that we have used many drugs to treat it, such as Sudafed and Claritn D to get through our summers. This has been a lifelong issue for all of us. Then one day while reading *The Power of Intention*, I had an "ah-ha" [moment]. I recalled from the movie the images of the brain producing all these chemicals, and Dr. Joe Dispenza, I believe, saying that the human brain has its own mini-pharmacy. So I decided to set my intention that every time I felt the effects of my hay fever start to kick in, I would visualize my brain producing what it needs to so that I could stop the effects of the hay fever on my body (like how the allergy commercials show what the drug does to block the allergy-causing particles). I have not used any external drugs to treat my hay fever in the last two months. It is almost a daily occurrence that I feel the effects start, but I have learned to stop it on my own.

In the beginning, it happened more often than it does now. I don't know if it is of my doing or the possibility that whatever causes my hay fever is not around as much, but I do know that these occurrences are dwindling away.

—NICK

A series of synchronistic events led to me seeing *What the BLEEP* here in Cardiff, South Wales. I have been reading books on quantum mechanics for the last two years (I'm no mathematician) and not knowing why I have been so drawn to them. I also have had wonderful experiences in the last year after visiting a crop circle at Silbury Hill, Wilts. Three months after that experience, I had a spontaneous healing. I had been diagnosed with macular degeneration of the eyes—a progressing condition that destroys the eyesight eventually. I was already experiencing distorted vision. I awoke one morning to find my eyesight restored. The specialist at my hospital was amazed, and he wants to make a special study case about me, as he said he had never seen this condition heal itself. After seeing *BLEEP*, I realized the wonderful potential within me, and it felt like a confirmation of something I had known all my life. Since then my eyesight has continued to improve, and I do not need such strong lenses to read. Day by day, I know I am creating my own circumstances.

—JENNIFER

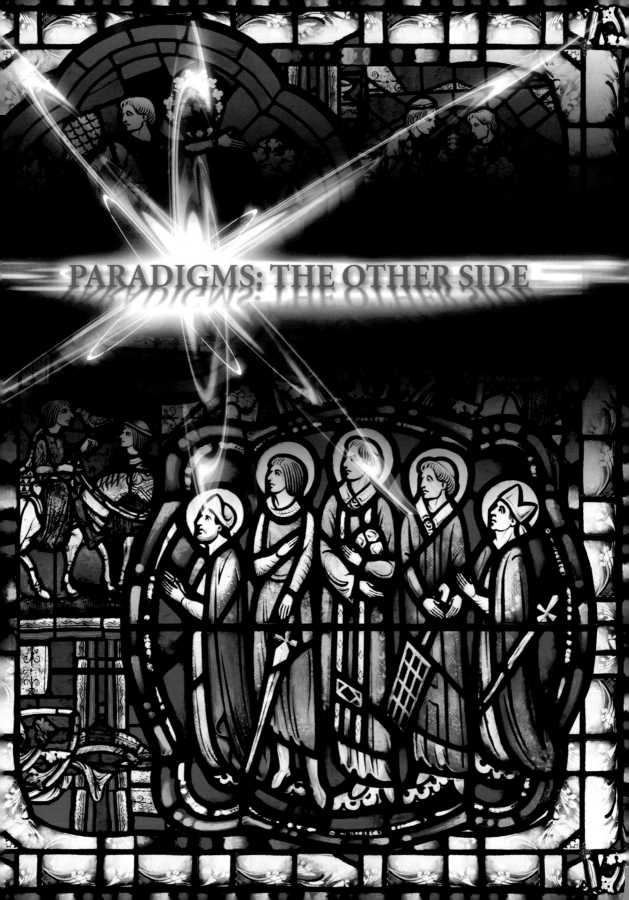
PARADIGMS: THE OTHER SIDE

A paradigm is a set of implicit
assumptions that are not meant to be tested;
in fact, they are essentially unconscious.
They are part of our modus operandi
as individuals, as scientists, or as a society.

—A Bunch of Chapters Ago . . .

We've been going on and on about the scientific paradigm and how it has determined attitudes and the way we look at the world. "Paradigm shift" is a much bandied about phrase that envisions a wonderful new world in which the old basic assumptions of the Newtonian worldview are gone.

But in the world of paradigms, there's an 800-pound gorilla standing right in the middle of the room.

Say hello to Religion.

And although many people want to say good-bye, religion in all its varied forms is the queen mother ship of paradigms.

"*. . . not meant to be tested . . .*"

"*. . . essentially unconscious . . .*"

In fact, it was, as we will soon see, the modus operandi that kicked off the Newtonian spirit/science split. Science didn't do that; the Church did.

A World of Religion

Undoubtedly, the biggest dividing line between Eastern and Western civilizations is caused by the beliefs surrounding God.

Not only have they defined the cultures to a huge degree, but the actual beliefs themselves are radically different. While the major Western religions are built around an all-powerful one and only God, the Eastern ones often have many gods, or in some like Buddhism and Taoism, the "G" word is never even used.

The Eastern religious traditions teach that God (or the Tao, Brahman, pure consciousness, the Void, etc.) is everywhere, and can best be experienced within one's own self. Indeed, the assertion is that God *is* one's own self—or perhaps it would be better to say, one's own self is God.

In the Western traditions, God is apart from us. And although Descartes is credited with "inventing" dualism, this dualistic idea of a God "out there" predated Descartes by thousands of years. In some sense this divided view of God/humanity opened the door for dualism to take hold in the West and led to a host of problems (and a host of wonderful discoveries!).

Because Western civilization is the most dominant these days and seems to be taking over the world, we will primarily be talking about Western religions. And although we will be going through some of the ideas and dogmas and pointing out problematic effects of these ideas, it's important not to forget the other side of that coin: Lives have been inspired, and hope and insight have been gleaned by millions through the centuries from the teachings of the various churches.

Evolution and the Human Enterprise

One thing about us humans; we're relentlessly pushing onward. Sometimes it's onward to what we're not sure, but nevertheless we move forward. Because in all living systems, that which doesn't move and change and evolve becomes stagnant and dies. Says Miceal Ledwith:

> The method in science is to produce a conjecture about
> reality, and then you set out hammer and tongs to disprove

THE MUSTARD SEED

So . . . a mustard seed.
This parable is a brilliant parable
for an age in which there was
no science. Jesus knew this
two thousand years ago.
How is he going to teach a bunch
of fishermen, tax collectors and
riders on asses about quantum
physics? What are we really?
We are even smaller than the
smallest space of time, from zero
point energy, to its recognition.
So how small are we?
We are point zero, and we
fluctuate off of that zero.
So he says to his disciples that
the Kingdom of Heaven is not
in the sky. And look, many will
say that it is here, it is there.
But in reality, a mustard seed,
the tiniest of seeds, is the
Kingdom of Heaven.
And what does that mean?
It means who we really are and
where we have our eternalness,
where we have always been and
always will be, and that self is
smaller than a mustard seed and
the Kingdom in which God lives
is smaller yet. And that, my
beloved entity, is consistent
with the quantum world that
makes up the large, slothlike
distractions called reality.

—Ramtha

it. In that process you'll be able to strip away the accretions and the peripherals and hopefully you get at a core that's hard and fast. But as you know, 90 percent of new knowledge is going to the garbage bin in a few years anyway, and that's the way it ought to be.

Now, the single exception to that in the field of human endeavor is religion, because religion says, "Oh, we have all the truth already right from day one," and therefore it can't ever change, therefore we can't set out to refute it, and because it is never set out to refute the peripheral aspects of its presuppositions, religion has become more and more irrelevant to the evolution of human thought.

But it wasn't always this way, even here in the West. The most obvious and well-known example of this is the teachings of Jesus. The God of the Old Testament, who was the Godus Operandus at Jesus' time, was a wrathful, vengeful, spiteful entity. One who seemed to thrill in killing off firstborns, humans, goats, cattle. Didn't matter, he was "a jealous God." Which is rather different from Christ's words that "God is love." As Dr. Ledwith describes it: "There was a quite noticeable shift in the thrust of the teachings of Jesus from, say, the teachings of the Old Testament. For instance, he said it was said to the men of old, quoting Moses: 'An eye for eye and a tooth for a tooth,' but I say to you, forgive your enemies."

And, in fact, much of Jesus' teachings seemed to stress getting away from the white-bearded, cloud-riding figure who dealt out judgment and wrath, to a more personal God, or moving away from the idea of heaven as a place, to heaven as a state of mind. As quoted earlier, "The Kingdom of God is within you."

In fact, the revolutionary aspects of Jesus' teaching are still reaching out to us over the centuries. According to Dr. Ledwith:

And he always stressed how important the thought was, even superior to the action that might follow the thought. So if I, for instance, have a thought that I want to murder my neighbor, but actually never follow it through, well, then obviously, it does make a difference to my neighbor that he wasn't killed, but in terms of the kingdom that I am constructing within myself, to put it that way, there is nevertheless no difference between the thought and the action that follows the thought, because in the quantum view of the world which we're more familiar with now, obviously the thought is the deciding factor, the thought is paramount.

But somewhere along the road to now, the Western religions got stuck. Books were deleted from the Bible, and books were edited. Differing views were ruthlessly stamped out and as the Church rose to might, all roads led *from* Rome.

And now there are hundreds of splinters of faiths, each one having the last word on truth. As author Lynne McTaggart puts it: "One of the problems with organized religions, though, is that there is this sense of separateness. That it's only good to be a Protestant or that people who are Catholics are the only people who know the way. And I think now our current understanding of quantum physics is this understanding of complete unity and so that we have to derive our spirituality from a sense of unity."

Of the deleted books, the one most scholars think is closest to the true teachings of Jesus and the early church is the Gospel of Thomas. This document was discovered in 1945 near the town of Nag Hammadi in Egypt.

Evolution and the Personal Enterprise

So given what we know about worldviews and how they constrain our view of reality, how do the prevalent views that have devolved from the great teachings affect us?

Dr. Ledwith relates this story:

I had an incident once in Australia some years ago. I was talking about the heaven and hell idea, and this young man of twenty years of age or so said to me:

"You know, that was really interesting, but, thank God, thank God, I am not religious."

They have never sinned;
they have never wronged.
Wrong, yes, in moral
confrontations with society,
but that's their adversity,
that's why they're here,
to wrong, to learn, search and
use the wisdom of that to
create yet greater dreams.
We have created all of this
for such as that. All people
are divine people.

—Ramtha

"Is that so?" I said, and he said:

"Yeah, I'm not religious. My parents had no belief, and I grew up without any beliefs at all."

And he had told me the previous day that his father had died when he was eight years old, so I said, "Do you mind me asking, where did they tell you your father went when he died?"

He said, "Well, they told me he went to heaven."

I said, "That sounds suspiciously like a religious view to me." And I went down through the twenty things or so, you know, belief in right and wrong, retribution for good and bad deeds, etc., and on every single one of them he passed with flying colors as a person with a profoundly religious outlook on reality. He was never in a church, synagogue or mosque in his life, but was a perfect religious person.

According to Dr. Ledwith, the single greatest obstacle to our evolution is the way our culture often views God—as a God sitting up somewhere "registering the scores on his laptop as to whether we perform according to his designs or whether we're offending him, as it's put, an absolutely outrageous idea. How could we offend God? How could it matter so much to him? How could it, above all, matter that he would find it so serious a situation that he could condemn us to an eternity of suffering? These are bizarre ideas."

And they are bizarre ideas: that in this vast universe, where there are more galaxies than grains of sand in all the oceans, that in that vastness, a group of people—well, *men* actually—on a small planet got the exclusive franchise for the pearly gate arches of heaven. And every other being in the universe will spend an eternity of suffering in hell. It's hard to imagine a more bizarre idea. And if that's the sort of God you believe in, you just have to wonder: How does that affect your view of the world?

And yet there are hundreds of millions of people who are taught these ideas of fear of doing wrong, fear of damnation and fear of living. As Ramtha observed: "It's the only planet in the Milky Way that has habitation, that is steeped in enormous subjugation of religion. You know why that is? Because people have set up right and wrong."

But this idea that somehow right and wrong mess up the true meaning and attainment of evolution is at first a hard one to accept. Is it okay to murder someone? Can I steal from my neighbor? Destroy a city?[1] The reason this is such a hard concept to grasp is because it is fundamentally a different way to approach life and evolution.

Right and wrong are based on a set of rules. These rules have evolved from teachings, cultural values and political convenience. They are all coming from outside, from our cultural beliefs. Evolution comes from inside. Looking at decisions from the perspective of evolution makes the basic assumption that, at the core, each person is basically divine. This is very different from the idea that we are born sinners or wretched beings who must be told exactly what to do because we are disgusting by nature. And thus, to keep these human animals in line, there is the threat of an eternity of suffering.

"Sinners in the Hands of an Angry God" is the title of a famous sermon by Jonathan Edwards, one of Puritan America's earliest fire-and-brimstone preachers. Now it could be argued that at certain stages of evolution the human race needed this type of "guidance," just like a three-year-old is told not to stick his fork in the electrical outlet. But what we're suggesting is that it is time to move beyond a right/wrong judgment-based system of values to one of personal evolution. And just like the child learns about higher forces—electricity—so the person learns about higher laws of the universe: "Judge not, and ye shall not be judged: condemn not, and ye shall not be condemned: forgive, and ye shall be forgiven" (Luke 6:37). Or, as we've been saying, your attitude inside will be reflected in the reality outside.

So I think anyone who is setting out on the path of enlightenment will be absolutely impeccable in everything that they do. Is it because of fear of damnation? No. Or of the punishment of God, or because I have sinned and haven't got forgiveness? No, no, no. The really enlightened person will see that every action has a reaction with which I must deal, and if I'm wise, I am not going to do stuff that will cause me to have to face it and resolve it and balance it in my soul later. That's the real criterion. So, you cannot sin against God because the divine presence is in us all, and we are fulfilling the divine mandate in everything that we do. We will find out quickly enough because of the action-reaction reality that there are some things that do not evolve me, and we'll have to learn that sooner or later. But, can you sin against God? That is impossible.

—Miceal Ledwith

---

[1] Of course, it's all scale. If a person does any of the above, it's "bad"; if a country does it to a neighbor, it's righteous conquest.

This concept of evolution based on increasing wisdom is much closer to the four noble truths of Buddha mentioned earlier: "The origin of suffering is attachment to transient things and the ignorance thereof." Suffering is not the consequence of displeasing an angry, jealous God; it is simply the result of ignorance—not knowing.

And as such is not a punishment, but a pointer, a nudge, a road sign on The Path.

Save Me from Myself!

One of the basic tenets of Christianity is the notion that "Jesus will save me." And, in fact, I have no hope of doing it myself, being a sinner born in sin and screwed from the get-go. It is difficult to imagine a more disempowering idea.[2]

The irony is, it's just those mistakes, ignorant decisions, "sins," that evolve us into higher and higher states. And if someone can always save you, then you never really have to take responsibility for anything, which is the classic victim mentality. In fact, many of these ideas have *victim* written all over them.

JZ Knight puts it this way: "Pleasing god gets us off of the hook of living. I mean, the very fact that we had to have somebody die for our sins is sort of a rip-off, don't you think? I mean, I think we all really should have had the privilege of living out our own sins and by virtue of experience being enriched with wisdom because of them. And I don't see how we can ever grow and become astounding beings unless we are fraught with the blight of experience that is bad and harmful and all those things, because only then have we really achieved the wisdom that allows us to understand everybody."

---

[2] If you can come up with one, you may want to add it to *The UnWizard's Handbook* section on turning Magicians into toads.

Is Religion *Bad*?

It's starting to sound like religion is the culprit—turning us into fearful sheep, awaiting the hammer of judgment to fall upon us. Furthermore, there is the unfortunate history of holy wars, witch hunts, book burnings and caste systems, to name a few. Not exactly the picture of divine guidance. But it would be rather hypocritical to slap the *bad* word on religion after going on and on about how good and bad don't exist.

Let's look on all the religious activities as a paradigm, or as a bubbling up of consciousness, that finds its expression in the various religions. And there is certainly a lot of data to support this. Look back at all those "bad" things: wars, witch hunts, book burnings, caste systems—*all* of them have had frequent expression outside of any church. If humans want to go to war, they'll find a reason. If people want to feel superior to others and that they have "the only way," they will find a paradigm that gives them that identity.

For a while some religions may have had a true divine source, but in time they are administered and devolved by humans. And so doctrines like "original sin," which was never part of Jesus' teachings, get tacked on. "Turn the other cheek" becomes "kill the Arabs in the holy land." But importantly, the root of these developments lies not in teachings, but in the recesses of humanity. The way out is not attacking religions, but evolving into the next generation, the next version of humanity. We don't get rid of wars by getting rid of churches or governments, but by evolving past those addictions that drive us to do what we do. The problem with religions is that they got stuck and dug their heels in about what is the truth. For the religious impulse to evolve past this stagnant stance, it needs to look at the other side of the coin—science—and learn what made science great. What's missing is the willingness to be wrong in the search for an ever-greater understanding.

Change happens. Certainly there are signs all around how

During the course of the last year while traveling all over the world talking to fans and press, I had a realization very similar to Will's understanding. I was using science to make fundamental religions wrong. Using a "might is right" approach, I was displaying the very attitude that I saw in some Calvinistic traditions. At the same time, I was hoping that these divergent traditions could find common ground. Perhaps they will only find common ground when I start doing the same. Trashing any one idea presupposes that there is only one right answer. And quantum mechanics has clearly shown this is not so. This has been my great humbling.

—MARK

When I was young we went to church every Sunday. The thing was, I never knew what church I would be going to. We went to a different one each week, mostly because my dad loved to hear the music. I was always amazed at how similar each church was in their understandings—love, kindness, unity. But I was also saddened by the separateness of it all. I would ask my dad—why so many different churches? He would say—because man has a habit of thinking he and only he knows the right and wrong, and those who don't agree with him are wrong! One day man will hopefully learn that there is no "right and wrong"; there are only degrees of understanding for which you must come to terms within yourself to evolve.

—BETSY

the images of God are shifting. The "Big White Father" on the throne of judgment, while very popular in the Middle Ages, is being replaced by a less human, more abstract idea.

Says Andrew Newberg: "People have kind of gotten away from God, the person, [to] having a more greater, infinite conception. God, in some senses, pervading the world, but also, in some senses being in addition to the world itself. So, it's not that God is the universe, per se, but that God basically emanates throughout the universe, and that would include all things and all particles and all beings."

This would include the idea of God that God is within.

The Sumerians saw God as beings who walked the Earth and interacted with them. Ed Mitchell's vision, as discussed earlier, was the opposite: "In one moment I realized that this universe is intelligent. Consciousness itself is what is fundamental." The Native Americans saw the spirit as everywhere and imbued in everything. Thus they would talk with the spirit of the wind, the spirit of the buffalo.

So what is God? What is Spirit? What is spiritual? Is spiritual just this realm that science got bumped out of 400 years ago? When's it coming back?

God/Spirit/Matter/Science

Is the divorce over? What's the settlement? Is *this* quantum divorce? Who gets the churches, and who gets the laboratories? Who cheated on whom? Who got cheated? Who's to blame?

Religions started it with ideas like a dualistic universe with God up there and us peons down here. In some aspects, the Church tried to take over all forms of human endeavor. Not just science, but all the arts, too. Until science proved them in error—so at that point the only solution was to divide up the pie. God and company get spirit; science gets matter. And isn't it ironic that this division is what drove science to the heights it is at today? If we look at Eastern religions, they never had the dualistic dogma take on things, so science was never excommunicated. And yet it was that extreme separation from the spirit world that enabled Western science to discover that mind and matter are the same thing.

What!? Well, if that's true, then the separation of science and spirit is truly artificial and not based on any reality. And who cheated on whom?

No one and everyone. These two worlds are two sides of the same coin, and that coin is humanity, flipping through reality. Maybe these two branches of human thought separated just so science could focus on matter and make discoveries and improve the quality of life. Certainly the dogmatism that is often pointed to in religion is equaled on the scientific side. It is like the varied religious forms themselves, simply a bubbling up of the underlying consciousness field of humanity. As John Hagelin previously observed: "Don't make the mistake of thinking that scientists are scientific. They're humans just like everyone else."

The Old Paradigm

Jeffrey Satinover said, "There are scientists who are as prejudicial as human beings as anybody else. There is the scientific method, which is specifically a method to minimize the influence of prejudice." Dr. Satinover goes even further; the split between science and spirit, he says, affects (or maybe infects) us all.

> The implication of the statement which arose during the Enlightenment is that every action that a human being takes, every event that occurs in life can be explained in principle ultimately in the way that every action in a game of billiards can be. The balls are all bumped at the beginning, and they all start moving around, and every single motion subsequent to that has absolutely no meaning. So

I think the important point to understand is that when we say God is within, we're not speaking of ourselves as a container within which God is enclosed. I always think of that old image of the alien movies with the creature bursting out of your chest; God is not like that within us. He's not encased within us. So when we say that the Kingdom of Heaven is within you, as Jesus put it, rather than God is within you, then we're talking about a complete identification of our being with the divine source that is at the origin of it all.

—Miceal Ledwith

anything you do in your life that you think is purposeful, intentional, meaningful, all of those thoughts about intention and purpose and meaning, it's all a complete illusion.

And so even people who don't believe it or understand it have been from their toes to the top of their head completely and thoroughly soaked and immersed and infected by that. And so, without realizing it—let me overstate this proposition a little bit—even people who absolutely insist that they don't believe it are utter believers in the mechanist view of life. It's impossible not to have been completely and totally infected by it. It dominates everything. And it has stripped us of almost any vivid sense that the world and the universe is actually alive. It's like a poison that has seeped into everything. Even people who insist they don't believe it are affected by it.

This sounds frighteningly similar to Dr. Ledwith's assertion that the religious worldview has seeped in, and we are all affected by it. And if both of these are true, that means that we are all living a schizophrenic worldview, a shattered disintegrated belief system: that our meaningless, purposeless actions are judged by some God, and although they are meaningless, our tour of duty in eternity depends entirely on them.

And who got cheated in this divorce? As always, the children. While the priests and the professors duked it out, humanity was left with a fractured confusing story of the universe. And both sides dug their heels in and shouted how very right they were.

Ask ten people whether or not they would like to know if there's a scientific basis to prayer and healing, thoughts and telepathy, and you'll get more yeas than nays. And still the vast majority of the scientific establishment will not take up the scientific method and dive into investigating these realms. Who loses? We do. Someone could have been saved if they had some assurance that their prayers actually helped, instead of being told by "those brilliant scientists" that it was all hogwash.

Scores of e-mails from people who watch the *BLEEP* film

have come in, and people were so inspired that intelligent scientists, with real credentials, were saying the "G" word (God) and talking about these things and, as reported earlier, had amazing things happen because they now thought these possibilities were real.

Dr. Radin adds: "If they are real, then it means that the large bulk of science has completely sidestepped something of profound interest and probably profound importance. Because anytime science has ignored something or partitioned off this area of the world, 'we are not going to look at,' it means that we are not getting a comprehensive view of reality."

And Now?

Yahoo! Humanity is bubbling again, and out of that pool comes scientists who are dedicating their lives to finding where these two seemingly disparate worlds intersect. Says Andrew Newberg:

> For a long time, science and religion really have been at odds with each other, and people have even gotten fairly adamant about one side or the other. What we've tried to do is take a different kind of perspective on the question, to find out how we can actually integrate the two, rather than separate them.
>
> And I think that when we do look at the relationship between science and religion, the neurological sciences [are] one of the richest areas for us to actually explore how these two can come together, because I think they can come together very successfully, without necessarily diminishing or even being combative with the other.
>
> So, what my goal has ultimately been with the work that we've done is to help to ultimately create a dialogue between people who are more religiously oriented, people who are more scientifically oriented, and say, you know

what? We can look at these questions in a safe way, in a way that preserves the science, but at the same time can be respectful and preserve the religious and spiritual implications of the science, and find a way to bring those two together.

Fred Alan Wolf adds: "It isn't a question of science bringing spirituality in. It's more a question of expanding the circle within which both science and spirituality lie, so that the kind of question we can ask can be looked at from the different points of view that both science and spirituality bring to the table. It's important to realize that the subject, the 'inner space,' is worthy of great exploration. It's important to realize that the ways we explore the 'inner space' may not be the same ways that we explore 'outer space.' But the ways that we understand inner space may be greatly assisted by the ways that we understand the quantum nature of the physical world."

The door swings both ways. Understanding the quantum nature of particles may be greatly assisted by including the study of consciousness. Certainly anytime two great branches of human endeavor come together, an entirely new civilization is created. Science itself is always seeking greater and greater models to describe the mysterious universe.

In many ways this intersection between God and matter, spirit and science, is summed up in Ramtha's answer to the Great Questions: "Why are we here? What is our purpose?" For it is indeed the rallying cry of science:

"To make known the unknown."

It was very important that, a long time ago, we made the decision to separate spirit from science. And so we were able to learn how to do science. But now we've learned, and we can take on the richer task of learning to do science when consciousness is part of the experiment.

—William Tiller, Ph.D.

Ponder These for a While . . .

- How has the religious paradigm affected your perception of reality?

- How is your paradigm a construct of your beliefs about right and wrong?

- What is right?

- What is wrong?

- Who holds the bag of truth on right and wrong? Do you? Does the Church? Do your parents? Your husband? Your wife? Does science?

- Do you see your paradigm expanding?

- Why are we here?

- Why are you here?

A Letter from an Observer

As a music supervisor, I'll watch a scene from the film hundreds of times while trying various pieces of music for that scene. Working on *What the BLEEP,* watching those scenes and the messages in those scenes, motivated [me] to address a lifelong fear of water and drowning, and to become a triathlete.

Joey Dispenza explains in one scene how, by thinking the same thoughts over and over, we cement connections in our brain's neural net. These repetitive thoughts cause us to act the same way over and over again. Joey then explained that, by changing our thoughts, we can "break" that neural connection and establish a new neural connection. With that change in our thoughts, and a sincere belief that we will become what these new thoughts describe, comes a change in action.

I decided to test this theory. I was scared of water and drowning (I literally could not get into a three-foot pool without that fear), but adding to this fear was that one of my brothers drowned in the ocean. I knew I wanted to become a swimmer. But it was only when I saw myself swimming, saw my movement in the pool and honestly and effortlessly believed I was that swimmer, that I started the work that swimmers do. The final stage was visualizing myself swimming a mile a day in the ocean. Starting in June 2004, and throughout the past year when conditions permitted, I averaged three to four miles of ocean swimming per week, winter included.

You have to *want it,* yes. But it's only when you effortlessly see yourself in a situation, and honestly believe that you are that person in that situation, that you act like you already are in that situation.

—TIM

ENTANGLEMENT

Nature, it appears, is made up as a nested hierarchy of non-locally connected, coherent systems.

ERVIN LASZLO

> We ran the generators, and sure enough we saw a spike with odds of a thousand to one.
>
> —Dean Radin,
> in regard to the O.J. experiment

Remember entanglement? It was Einstein's attempt to discredit quantum theory with the "spooky action at a distance"—the phenomenon in which two entangled particles can be separated to opposite sides of the cosmos, and something can be done to one, wherein the other responds instantly. It's sometimes called "non-local" because the idea of something being local means that it is not distant, and in entanglement there seems to be no such thing as distance. Everything touches all of the time.

Erwin Schrödinger was quoted earlier as saying, "Entanglement is not a property of quantum; it's *the* property." But that is quantum physics—the physics of matter, energy and particles. What about other areas of experience? Is this phenomenon seen in biological, societal or global systems? Or is the extrapolation to those areas just wishful thinking by New Age philosophers?

In many ways, these theories, experiments and debates are at ground zero of "The New Paradigm," and in essence draw the line very clearly between a dead disconnected universe and an intrinsically alive, interconnected one.

Entangled Minds

It doesn't seem too much of a stretch (to us, at least) to think that minds get entangled. Particles do, and particles are like information, and mind as matter, and matter as mind, so why wouldn't minds get entangled? Although the experiments on particles do not prove that minds get entangled, they certainly point to a compelling area of study, and one that any mind looking at a more general, inclusive theory of everything would want to investigate.

At least that's what Dr. Radin thought. *Entangled Minds*[1] seemed to answer a lot of anomalies we find in the world, so he decided to put it to the test in a lab. He began by asking two people to keep each other in mind throughout the whole experiment. Dr. Radin has found that he can entangle minds by simply asking people to keep another in mind. After this, the two subjects were separated and sent to two specifically different places so that no physical communication could occur between them. The scientists wired them both physiologically and did the scientific equivalent of what Dr. Radin calls "poke one of them and see if the other one flinches. And if it turns out that you can poke that person and the person flinches as a result, that is a way of demonstrating that they are still entangled even though they are no longer in the same place . . . This kind of experiment has been done with many different physiological parameters."

He does experiments like using a flashlight to shine light in one person's eyes and then seeing if the other person's brain, specifically the back of the brain called the exhibitor lobe, registers a change. He has found in experiments like this done over the period of two decades that "shining the light in a person's brain gives a very characteristic brain response in the person who has seen the light . . . The person who is sitting in a dark

---

[1] *Entangled Minds* is the title of his book on this subject. The experiments mentioned in this chapter are covered in great depth in that book.

I think there's a lot of evidence suggesting that a "disturbance in the force" is real; it's not just a metaphor, it is not mythology. Some people can feel it better than others, but yes, I would call that force this entangled mind thing [or] whatever words we want to use for it.

—Dean Radin, Ph.D.

room doing nothing, the partner, their brain does not light up in the same way because it is not being driven by sensory input, but it does change; it changes its behavior in a way that is more or less in time synch with what is going on in the sending side . . . Under an Entangled Minds model they are connected all the time; when I poke one, the other one flinches; it's not because something magically traveled over, but because poking one is like poking the other one, and that's why they flinch."

Dr. Radin reports the incredible phenomenon of this happening in terms of odds: "In research involving Entangled Minds, the evidence is about one thousand to one against chance for the kinds of connections that we see. This is based on a meta-analysis. In further experiments done on the common phenomenon of having the sense or feeling as though one is being stared at, the number according to chance has come up way, way above a trillion to one." That huge number is based on the collection of thousands of experiments carried out over decades and combined statistically into Dr. Radin's meta-analysis.

So is ESP (extrasensory perception) some sort of spooky action? Dr. Radin finds that by looking at the entire range of ESP phenomena as different applications of entanglement, they are brought under a unified theory. "Let's assume that experience is entangled, then how would it manifest? And we can start going through ways in which it would manifest. If there's a connection with other minds, call it *telepathy*; if there's a connection to some other object somewhere else, call it *clairvoyance*; if there's a connection that happens that transcends time, we call it *precognition*. If there's a connection in which my intention is expressed out in the world somewhere, you might call it *psychokinesis* or *distant healing*. So you can go through a list of perhaps twelve kinds of psychic experience that have gotten labels over the years like telepathy, but this is really just the tip of the iceberg."

How deep does the iceberg go!? If there's some sort of nonphysical entanglement phenomenon happening between two people, implying that there is some sort of mental space or mental field where interactions occur, there should be aggregate effects, or effects that scale up to hundreds, thousands and millions of minds. To see whether or not this is so, Dr. Radin and colleagues have inaugurated the Global Consciousness Project. Inspired by the results from the O.J. Simpson trial and the effect on random event generators (REGs), they have set up a worldwide network of REGs, which continually upload their readings to a server in Princeton.

Says Dr. Radin: "So we have data now for two kinds of events. We have planned events, like Y2K. We also have spontaneous events, like 9/11, that weren't planned. We're able to look at all of them now in terms of what happened to this randomness, this physical measure of randomness around the world when a lot of minds suddenly became focused on something, and the short story of this is—and we have hundreds of events now both planned and unplanned—and it's very clear that overall the randomness is not as random as it ought to be, by theory, when these events are occurring. Large-scale events that attract a lot of attention create a certain mental coherence, which seems to be reflected in what's going on in random generators around the world."

This is stunning news. The theory that Dr. Radin alludes to above is at the basis of quantum theory—the randomness of quantum events. Yet a coherent focusing of millions of minds changes it. And just as the discovery of the discreet quantum nature of particles poked holes in Newtonian assumptions, so this "ugly fact" ought to do the same for the mind/matter barrier that has characterized thought for centuries. The implications are wonderfully monstrous!

Entangled Everything and Complexity

Psychology, sociology, biology, economics, parapsychology, medicine, politics, ecology, systems theory, ethics, morality, theology—*all* are revolutionized by the concept that something non-physical—the mind, consciousness—has a real effect on the physical (making the mind as real as matter) *and* that coherent minds, minds thinking the same thought, aggregate into something that is also real. Real enough to drastically change what quantum "laws" predict.

It is at this point that we run into Complex System Theory. This relatively new branch of science looks at complex systems and how they are constructed from smaller, less complex systems. This theory is built upon the notion of "feedback loops" among all the different levels of complexity. In other words, the smaller structures affect the large, and the large affect the small.

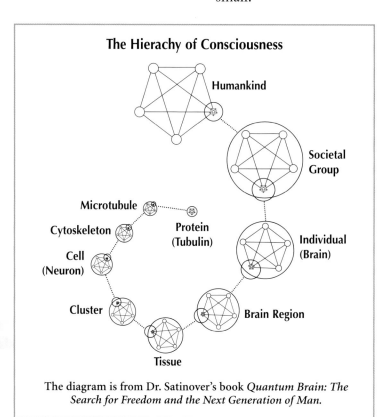

The Hierachy of Consciousness

The diagram is from Dr. Satinover's book *Quantum Brain: The Search for Freedom and the Next Generation of Man.*

Dr. Satinover used this theory in constructing his Quantum Brain theory as "a nested hierarchy of parallel self-organizing computational structures, characterized by chaos, bifurcation, sensitivity to initial conditions, 'spin-glass' dynamics, etc. At the lowest level are quantum effects. Because of the nested, hierarchical structure, iterative feedback of initially wholly indeterminate states, rather than being averaged out, are actually amplified upward through the various scales and generate large-scale indeterminacies."

However, Dr. Satinover maintains that the model, and thus the quantum indeterminacies, continues to scale up from individual brains to groups, societies and even the planet. And that this scaling need not take into account any mind-field or metaphysical effects to produce a form of coherence at the level of all of humanity.

In Complexity Theory, there is the concept of self-organizing structures. This property, sometimes called an emergent property, comes about as the system evolves into greater complexity. (Materialists use this concept to explain consciousness as emerging out of complex neuronets.)

In its current state, mainstream Complexity Theory does not include the Entangled Minds aspect. However, the two are tantalizingly close. They both address the issue of how things may scale up and produce something greater than the sum of the parts. The big difference is that with entanglement, the element of the spirit, or of the mind, is included as a real processing element, and thus affects the whole.

Now, let's stroll down a few of the "ologys" listed above and see how these theories are affected. For our discussion, we are accepting Dr. Radin's experimental data that point to entanglement of minds as a basic feature of reality. The following is really a brief overview of some of the interesting scientific developments along these lines.

The combination of these ideas—Entangled Minds and Complexity Theory—and the application of them to established disciplines has to be the most exciting scientific frontier of the age. We hope this whets your appetite for delving in further. For a complete explanation, check out the scientists mentioned directly.

I am reminded constantly of Gandhi's statement, "Be the change you want to see in the world." For many this is an elusive idea. The notion that somehow *me* changing is going to affect the rest of reality seems, well . . . like a sweet idea. But when you look at the work that Radin, Sheldrake, Hagelin or Laszlo are doing, Gandhi's statement is starting to make more sense. Why should we be the change we want to see? Because we are entangled with everyone. Every thought matters. Who we are "matters" to everyone else. What if it really is that simple?

—MARK

Psychology

Carl Jung, one of the founders of modern psychology, posited "the collective unconscious." It was from this collective unconscious that certain ideas or archetypes ("home," "child", "God," "hero," "saint") arise, eventually making their way into our conscious behavior. Jung noted how these archetypes seemed to be universal among humans, and give rise to similar concepts, dreams and other phenomena in human psychology. Many

It's a very interesting question about whether we have thoughts that we can actually control completely, or if an unpure thought pops into our mind . . . did we have control over that? And should that be considered to be immoral, because that happened? So there's some fascinating ethical and moral questions that come up because of this kind of research. I think we need to explore the questions in more detail.

—Andrew Newberg, M.D.

techniques and therapies were developed based on this idea that have been used successfully by Jungian therapists for decades.

What the idea of Entangled Everything says is that this theoretical construct has a very real basis in reality. It is not just some nebulous fluffy idea to help deal with human psychology, but that Jung deduced and/or intuited something of the structure of the universe, and it's taken nearly a hundred years to back that finding. It's no different than physicists deducing the existence of quarks, and then finding those particles decades later.

What this means to you is that you are affected by what everyone else is thinking, feeling and experiencing. Walking into a room where everyone is rattling their swords, itching to go to war, will affect you, *and* you walking into that room affects them.

Sociology

"Social consciousness" is not a phrase that just describes a certain herd instinct, but that herd instinct is itself as real as the book in your hands. And those social agreements are created by many minds being coherent around a given subject. The "bubbling up" of paradigms discussed last chapter is another expression of the aggregating affect of coherent minds.

John Hagelin has been pursuing and studying these effects for years. He provides a number of statistical studies showing how small focused groups affect society at large: "The spillover effect of that individual enlightenment is going to change the surrounding society more powerfully than we might think. And that's the peace research I alluded to briefly in the film, and there's been more research since to show how powerful groups of individuals can be collectively meditating and stimulating the unifying field at the source of mind and matter, creating indomitable waves of peace and unity in the world."

Biology

With the discovery of DNA, the dream of finally understanding how life is created and sustained seemed within reach. The DNA was seen as the computer code of life, and once decoded would show how life truly works. Since then, scientists have found that there isn't enough information in DNA to describe how to create a body from a fertilized egg.

Rupert Sheldrake is a biologist who began investigating the anomalies of biologic systems. In response he has developed: "The Hypothesis of Formative Causation [which] states that the forms of self-organizing systems are shaped by morphic fields. Morphic fields organize atoms, molecules, crystals, organelles, cells, tissues, organs, organisms, societies, ecosystems, planetary systems, solar systems, galaxies. In other words, they organize systems at all levels of complexity, and are the basis for the wholeness that we observe in nature, which is more than the sum of the parts."

Sheldrake has investigated such strange effects as how pets seem to know when their owners are coming home, and the common sense of being stared at. His theory is too broad to be discussed here, but the following statement applies to the "entangled everything" approach:

> In discussions with the late David Bohm, it became clear that some of the phenomena I am talking about in terms of morphic resonance and formative causation could perhaps be explained in terms of non-locality in quantum physics. Further discussions of non-locality in quantum physics have led me to think that a new theoretical framework should be possible within which quantum non-locality and morphic fields can be integrated.
>
> I do not think that the quantum physics of subatomic particles can be directly extrapolated to account for the morphic fields of living organisms. After all, even existing quantum physics is hard to extrapolate to complex molecules

We have to bow to a greater mind that is forming this energy into modes of reality that we have yet to dream in our lifetime, and we only perceive it yet as chaos, but its order is definite. It's above us. It's deeper.

—Ramtha

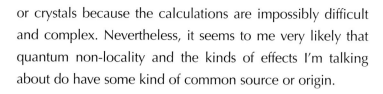

I think that when two people are talking, they may be communicating on two different levels. They may be communicating on a classical level, and there may also be some kind of non-local interactions, due to some kind of quantum entanglement, quantum coherence between the people. Political scientists are getting interested in this now to look at the implications of quantum interactions between people, societies and even governments and how that affects politics.

—Stuart Hameroff, M.D.

or crystals because the calculations are impossibly difficult and complex. Nevertheless, it seems to me very likely that quantum non-locality and the kinds of effects I'm talking about do have some kind of common source or origin.

What this is suggesting is that Sheldrake's morphic fields are the crucial factor in deciding which emergent property actually emerges in complex systems and that the morphic field is a field of coherence with its source in something non-physical—which is *real*.

These morphic fields can be thought of as the blueprint, or template, which when impressed upon the random nature of quantum events, changes those events into some of a higher order. It is like an invisible hand guiding which property of many possible ones emerge into the physical world.

Politics

John Hagelin has taken the societal repercussions one step further: "We've established what's called the U.S. Peace Government, not to compete with the existing government, which is primarily concerned with the management of crisis, but a government of the people that is designed to prevent problems from arising through education that raises the collective consciousness and elevates human behavior to be more in harmony with natural law, promoting sustainable solutions in the areas of agriculture, energy, education and crime prevention."

This peace government would not function by opposing the existing government, but doing what Dr. Hagelin calls "an end around." By utilizing the knowledge about morphic fields, the collective power of coherent minds, and implementing an enlightenment-based education system, he foresees an evolutionary shift in the way the world is governed. Certainly, if the entangled minds hypothesis is correct, there are a plethora of applications for those who understand the implications.

Ecology

One of the first theories, which looked at a holistic approach to physical phenomena, is the Gaia Theory. In the early '60s, James Lovelock was working for NASA in its search for life on Mars. As a result, he started looking at what constitutes life on Earth. He began to realize that everything about the Earth's environment was interrelated and, in fact, regulated by inorganic and organic factors. Dr. Lovelock remembers:

> For me, the personal revelation of Gaia came quite suddenly—like a flash of enlightenment . . . I was talking with a colleague, Dian Hitchcock, about a paper we were preparing . . . It was at that moment that I glimpsed Gaia. An awesome thought came to me. The Earth's atmosphere was an extraordinary and unstable mixture of gases, yet I knew that it was constant in composition over quite long periods of time. Could it be that life on Earth not only made the atmosphere, but also regulated it—keeping it at a constant composition, and at a level favorable for organisms?

The idea now of the planet as a great living being, with rivers for circulation, for example, is a common thought, and has spurred much of the interest in ecology. But Lovelock and proponents of the theory received criticism that they were implying there was a purpose behind nature's regulation:

> Teleology, from the Greek word *telos* (purpose), asserts that there is an element of purpose or design behind the workings of nature. It is part of a very old debate between mechanists who believe that nature essentially behaves like a machine, and vitalists who believe there is a non-causal life force. Critics thought Lovelock was saying that the planet had a life force, which was actively controlling the climate and so on. However, this wasn't Lovelock's intention. He stated that, "Neither Lynn Margulis nor I have ever proposed that planetary self-regulation is purposeful . . . Yet we have

If I'm always connected or entangled, how do I know it? Is there something there I should be listening to? Will it guide me where I want or need to go? I've begun to realize it's the subtleties that are so small (I barely notice them); they are actually the big clues. It's about being observant enough to see them in everyday things. This is how I know I'm connected. Today my daughter grabbed a picture of my friend and her baby off my fridge and began to babble to them. Moments later the phone rang and sure enough—it was my friend.

—BETSY

met persistent, almost dogmatic, criticism that our hypothesis is teleological."

It seems that this debate has been raging for quite awhile. The Entangled Minds theory brings the curtain down on the discussion by saying that the life force is the morphic field, which is an aggregate of minds large and small on the viability of life itself. It says life is not just random mutations, but emerges from an ever evolving, non-physical source. Consciousness creating reality.

The Connectivity Hypothesis

The scope of implications of this new model is nearly limitless. Although we meandered through a few corollaries to it, the scope of the changes to existing thought are more far reaching than any previous scientific discoveries. It's not a paradigm shift, but more like a paradigm meltdown—with the paradigm as a phoenix. Burned to ashes and rising, or emerging, anew.

And the scope of this has not gone unnoticed in the scientific community. Divergent fields of interest such as biology, physics and sociology are for the first time exchanging ideas that move those sciences ahead. It is the dream of many scientists to come up with the one all-encompassing theory of everything. Dr. Ervin Laszlo states: "Einstein's pronouncement, 'we are seeking for the simplest possible scheme of thought that can tie together the observed facts,' is the aspiration as well as inspiration of Connectivity Hypothesis."

Dr. Laszlo is widely regarded as the founder of systems philosophy and general evolution theory. His Connectivity Hypothesis is a unified, concise theory that takes many of the discoveries we've covered and puts them into a conceptual framework that forms the "Foundations of an Integral Science of Quantum, Cosmos, Life and Consciousness."

In his books, Dr. Laszlo discusses anomalies from the worlds of medicine, biology, parapsychology and physics and introduces

Matter as well as mind evolved out of a common cosmic womb: the energy-field of the quantum vacuum. The interaction of our mind and consciousness with the quantum vacuum links us with other minds around us, as well as with the biosphere of the planet. It "opens" our mind to society, nature and the universe.

—Ervin Laszlo, Ph.D.

a mathematical formalism that makes testing of his hypothesis experimentally practical. One of the cornerstones to this theory is the role of information, or as he puts it, in-formation:

> The presence of in-formation in nature is a fundamental tenet of Connectivity Hypothesis. The hypothesis rests on three laws of in-formation as a physically active element in the universe:
>
> i. changed particles and systems constituted of charged particles create physically active information;
> ii. the information is conserved;
> iii. the information created and conserved feeds back to [or "informs"] charged particles and systems of particles.
>
> Connectivity Hypothesis shows that the feedback of active information, occurring in the holographic mode, creates coherence among the particles and systems of particles that created it.

Those systems of particles can be electrons, photons and molecules or cells, humans and civilizations. And the information that in-forms them is as physically real as those particles: mind as matter. In this Dr. Laszlo has given a scientific formalism to such diverse concepts as collective unconscious, morphic fields and entanglement—both particles and minds.

Dr. Laszlo concludes his book with the following provocative statement about what the hypothesis points to, and the answer to "Why are we here?" on the grand cosmic scale: "The evolution of the metaverse through cyclic evolution of universes conduces to the full realization of the evolutionary potentials encoded in the primordial cosmic plenum[2]—to the complete coherence of all things that exist in space and time. It marks the full achievement of divine creativity: ultimate coherence in the mind of God."

---

[2]Cosmic plenum, according to Dr. Laszlo, is the "misleadingly named quantum vacuum."

UP AND DOWN THE RABBIT HOLE

And what this gives is a mechanism by which mind, or consciousness, creates structures or reality out of smaller pieces— consciousness creating reality. And if minds are also elements in a complex system, they give rise to greater minds that have properties, godlike properties, that emerge out of the chaos of complexity, that scale all the way back to—God? Point Zero? the Void? Which because of the feedback structure of these systems, patterns our reality "down here." Which due to the feedback system, feeds up . . . How far *round* do you want to go?

—WILL

Ultimate Coherence

What about ultimate coherence in our minds? Is this what we unconsciously strive for? Is this why we gravitate to a thousand voices raised in song, or why a beautiful rainbow with a loved one is far better than seeing it alone?

There have been times on this planet when humanity got together for a greater good, and the feeling of goodwill that was generated was overwhelming. Dr. Emoto, who has been all over the world creating coherence around the concept of water, made this observation about the *Apollo 13* crisis: "At that time, because of the spread of the TV by then, many people in the world, myself included, prayed for the safe return of the three astronauts. And that was true for even Muslim people, Christian, Buddhist, Jewish, it did not matter what religion, race or ethnicity one was. That's why I believe that such a miracle occurred."

During the disaster of the tsunami of Christmas 2004, a similar mass consciousness occurred. Dr. Emoto continues:

> When the people around the world saw the images of the destruction that has killed over 200,000 people, I think they realized that that could have easily happened to themselves. They were able to think of this incident in a more personal perspective. Through my ten years of research, I have been visually sending the message to people all over the world that people's prayers and thoughts can affect reality no matter where they are. So because this idea had already been accepted by many in the world, people probably thought, 'Oh, so that's how it works. If we try, our energy can reach there. Then we should all pray.' And this is the reason why there are no great outbreaks of infectious disease as a result of the tsunami disaster. With a repetition of similar actions, people may start regaining faith in prayers.
>
> Therefore, in other words, there might be more disasters, be it natural or man-made, that would create more casualties along the way, but through these experiences more

We are all connected. We are entangled; if you want to call it quantum entanglement, fine. But we are entangled. And there is no real separation between us, so that what we do to another, we do to an aspect of our self. None of us are innocent in that regard. There's something out there we don't like; we can't really turn our backs on it because we're co-creators, somehow or another. And we have to do the right things to try to get the future that is best for all of us. That's our responsibility as co-creators. And in the process, whether we become a politician, whether we become a theologian, whether we become a scientist or a doctor or whatever, we all can contribute to life and lift it to the utmost of our ability and to do the things that we think are the best things to do. Which require that we really think about things. We reflect and we act, recognizing that all the others are our brothers and sisters, and it's a family matter. That's it.

—William Tiller, Ph.D.

people may be able to perceive these problems as their own, that they are all responsible as well.

I don't know about you, but those few times when humanity acted globally to help humanity, the feelings and effects were quite literally out of this world. Or rather, out of the "more for me, mine is better than yours" world, and into a high-as-kites happy world of global-Gaia coherence. We know coherence does something. It somehow, some way pushes random quantum events around.

Coherent *intent* does something even more, to borrow Dr. Laszlo's words: "It marks the full achievement of divine creativity."

Ponder These for a While . . . .

- How would the Entangled Minds/Connectivity Hypothesis change the following areas of study?
    - Economics
    - Parapsychology
    - Medicine
    - Politics
    - Ethics
    - Theology
    - Education

- What would this world be like?

THE FINAL SUPERPOSITION

I'm now running through a list of famous people who have claimed
that something is the final word, and I can't think of
a single one who hasn't been proven wrong.

JEFFREY SATINOVER

> **There is superposition of multiple possibilities, which after a while will collapse to one or the other, so you choose to do this or choose to do that.**
>
> —Stuart Hameroff

> All of the concepts in theoretical physics, the concepts on quantum physics, all that's fabulous and wonderful. But at the end of the day, does it give us a better way to change our mind?
>
> —JZ Knight

Lest we end up on Dr. Satinover's list, let's say from the start that this chapter is not going to be a final word on how to run *your* life, or what *you* need to do, give up, do without or create. As if we would know, sitting here typing at a computer, what crucial thing, place or event will turn your life around and make glorious your trip through the universe. People often seem to think that there's a secret formula, that if only they would be told, then everything would be wonderful. If only they could get five minutes alone with Donald Trump, then their business would fly.

So then, why have we been writing about all these ideas, experiments, concepts, methodologies and ways of seeing the world? Think of it as loading up the tool belt. And we're loaded with some *heavy* equipment.

Many of the concepts we've been talking about have been debated, discussed and argued by the best minds and spirits on the planet for thousands and thousands of years. So in the realm of mind (which is real, after all) we've joined with many truly amazing beings. You might ask: How can we even hope to definitively unravel what these geniuses could not? Well, they probably thought the same, but went into the unknown anyway. And in the end, the answers to What Is Reality? Who Am I? Do I Create Reality? What Is Matter? How Do I Become

Enlightened? What Paradigm Am I Stuck In? are intensely personal anyway. We have to answer them for ourselves.

Okay, but another question immediately comes to mind—why worry about such "lofty" ideas when you hate your commute to work? Why wonder what is reality, when you are "stuck" in yours? Dean Radin was asked why he cares about philosophy and abstract thought:

> Because it goes to the essence of the assumptions of who and what you think you are. So if we think that we are living in a certain kind of world, we behave in a certain kind of way. If we think we are living in a world where human beings are a kind of machine, we're like robots walking around, and there may or may not be anything actually happening in there, then questions like morals and ethics and how we live our lives and what we think about death and life, those will be very different than if we think the world is an interconnected, alive one.

Why Care About Science?

One of the first reasons goes right back to what science is based on: the scientific method. Given that what we perceive is based on what we know and what we believe, it seems difficult to ever get a true picture of the way things really are. The scientific method is a revolutionary approach to reality; as far as possible, it removes the prejudices from the observer and thus gets a truer picture of reality. The reason why that is important can be seen by reflecting on the Middle Ages, where the conception was: The world is flat, and there is a edge to fall off of. This was hardly the basis for an age of exploration. So people by and large stayed on the farm, in the town, or in the fiefdom.

In other words, our understanding of reality limits our options. The grand thing about science and its method is that it has the ability to say, "What I thought was reality was only an

Well, science creates the stories that we live by, and science has told us a very bleak story for the last hundred, four hundred years. It's told us that we are some sort of genetic mistake. That we have genes that just use us, basically, to move on to the next generation, and that we randomly mutate. It's said that we are outside of our universe; that we are alone, that we are separate. And that we are sort of this lonely mistake, on a lonely planet, in a lonely universe. And that informs our view of the world. It forms our view of ourselves, and we are now realizing that this view, this view of separateness, is one of the most destructive things. It's the thing that creates everything; all the problems in the world, the wars, the views of I need more than you, the aggressiveness in everything from business to the classroom. And we're now realizing that paradigm is wrong. That we aren't separate, that we aren't all alone. We are all together. That at the very nethermost element of our being, we are one; we are connected. And so we are trying to understand and absorb what are the implications of that.

—Lynne McTaggart

approximation—I now have a better one." Think about what kind of a tool *that* is in the old belt.

Not that science is the only way to approach life. There is art and beauty and inspiration and revelation. Nevertheless, think of all the times you didn't do something because you might be wrong or fail. In science there is no such thing as a failed experiment. That experiment was successful—it told you that reality does not work that way. Why care about science? We asked John Hagelin:

> I want to stress emphatically that everything I am talking about here really represents a firm foundation of mathematical physics with predictable consequences that can be tested in a laboratory and more importantly applied for the benefit of society. That's what the ultimate importance is, the discovery of the unified field; the superstring is beautiful, but more importantly the discovery of the unified field will soon transform civilization away from today's fragmented world, which is crisscrossed by arbitrary political borders that separate humanity from humanity.
>
> A fragmented world reflects a fragmented understanding of the universe. Now with the emergence of the fundamental understanding of the unity at the basis of life's diversity, it won't be long before this rainbow-colored, politically divided world will become a global country, a global country of peace. And that we are going to achieve it in our generation.

Simply put, science tells us what is possible. People tend not to venture into what they believe to be impossible. But what is impossible? Quantum theory says that it is possible, in the next instant, for you, your body and the chair you're in to, for no apparent reason, be on the other side of the universe. The probability is ten to the minus gazillion, but it's not zero.

So Candace Pert says: "The body always wants to heal itself. There is a database of spontaneous remissions and spontaneous recoveries, particularly from cancer, and what's interesting to me is that it's so often accompanied by a sudden release of emotions." Then somebody reads it, releases something and is healed.

Why Care About Change?

Really. It's a pain in the butt. Everybody and everything starts screaming. Bosses, lovers, parents, why, even your cells are clambering for that good, old feeling. Just slip out of that tool belt and onto some comfy couch. Or not. Says Dr. Joe Dispenza:

People have to make that choice for themselves. Most people are happy with their life the way it is. Most people are happy with watching television and having a 9 to 5 job. Not to say that they are happy with it, but they are hypnotized into thinking that that's normal. The person who has another urge inside of them that they're clearly interested in something else, all they need is a little bit of knowledge, and if they accept that knowledge as a possibility and if they embrace that knowledge over and over again, sooner or later, they'll begin to apply that knowledge.

Now for some people it may take five minutes, and for other people to take that first step may take an enormous amount of effort because they have to weigh that first step against everything they know, and everything they know is attached to the way their life is presently, all their agreements, all their relationships. And to take that first step means that they have to evaluate what it's going to look like by taking this step against what they know, and there's that battle between those two elements. But once we give ourselves permission to move outside the box, there's a definite sense of relief and definite sense of joy.

What is the definition of any miracle? Something that happens outside of convention, outside the box of what's socially acceptable, scientifically acceptable, religiously acceptable. And right outside the box is where human potential exists. How do we get there? We have to overcome the emotional states that we live in on an everyday basis. Our own personal doubt. Our own unworthiness. Our own lethargy and fatigue. Our own voices that say, we're not good enough or it's impossible.

—Joe Dispenza

How can we say that
we have lived fully every day
by simply experiencing the
same emotions that we're
addicted to every day?
What we're actually saying is,
I have to reconfirm who I am,
and my personality is,
I have to do this, I have to go
here, I have to be that.
A master is quite a different cat.
It is one that sees the day as an
opportunity in time to create
avenues of reality and
emotions that are unborn,
of realities that are unborn,
that the day becomes a
fertilization of infinite
tomorrows.

—Ramtha

So according to Dr. Dispenza, on the other side of "another urge" is a "definite sense of joy." Really? At this point we could launch into all the reasons why, and talk about evolution and "making known the unknown," but that's what we've done throughout the book.

And besides, What the BLEEP Do wΣ (k)πow!? anyway? Dr. Dispenza obviously thinks it's true, but that's not me, and that's not you. The real question is: What the BLEEP Do YOU (k)πow!?

Really.

The short-form answer to that is: Try it. Test it out. Test it all out. Remember, the answers are intensely personal anyway, which is why we're not giving you the quantum cookbook on how to cook up a wonderful life. The good news and the bad news is that only you know.

But once the process of experimenting with your life is engaged, all the information we've been going over and all the knowledge about creating and emotions and addictions, about choices, changes, intent, about associative memory, will all come into play. The paradigms and neuronets that limit your degrees of freedom will leap up in your face. The beliefs that constrain your immortal spirit will scream in your ear. It'll be chaos, mayhem. Whoa!—you'll be alive.

The Final Collapse

The final collapse to the final superposition is you. All these possibilities: to change or not to change? To change into what? To own, retire or collapse what emotional pattern into wisdom? Which belief to put to the test and find out what is real? All these possibilities are sitting out there, awaiting a choice. Says Miceal Ledwith:

What's the difference between belief and knowledge? Well, I believe something on the authority of some other person or thing. I have knowledge when I have experienced that thing personally, and if I, for instance, walked on water, I would have knowledge that it is possible to do, and I could never again doubt it really. But if I only believe it on the word of somebody else, well, then it's only a philosophy, an abstraction, and a great necessity in evolution is to change belief into knowledge or into experience or into wisdom. To convert knowledge into wisdom that is experienced is the great journey of spiritual development.

And we are apparently equipped to make that journey. Dr. Dispenza adds: "The brain is actually a laboratory, and by our design, and by our own will, it acts as a laboratory, to take concepts, ideas and models, and ask the what-ifs, the possibilities, the potentials, and contemplate on designs or ideals that are outside the boxes of what our present understanding is, to come up with a new understanding or an enlarged box."

As they said in *Ghostbusters*: "We got the tools; we got the talent."

And you just gotta wonder, why do we have these tools? These talents? Either it's an accident of nature or it's why we're here. It's pretty much one or the other. Obviously, the thrust of this book has been on the "it's why we're here" side. All the creations of humanity spring out of the abilities, the human potentials, that we are in possession of. And we have them for a reason, which we're all in the process of discovering.

We have the amazing brain—the most complex structure in the known universe—that can and does rewire itself to continually maximize whatever you want to experience. Whereupon the brain rewires in response to that new experience—all under your control. Then there's the body: self-healing, self-replicating, and let's face it, a thing of beauty. And the mind, which has the ability to delve into the tiniest corners of space and time and then get huge and contemplate the big bang. And beyond.

And within. Consciousness explores consciousness and comes up with such crazy ideas as: The world is essentially empty; all we perceive is maya—illusion—and we are

Enlightenment is our birthright. We're wired for it. It's what the human brain was designed to experience.

—John Hagelin

fundamentally all connected—we are one. Those explorers of the unseen, the enlightened, have been reporting on this for millennia and lo and behold, out of those tiniest corners of space and time, and out of the probings into the working of the brain, comes the message, "Yeah, that is what is."

The tool belt is getting very full. We've always had the tools for transformation; all that's missing is hitting the "go" button. If there's knowledge that we're without, we'll find it; if it's an experience we're missing, we'll create it.

Higher and higher, on and on. Remember those nested Chinese boxes? How properties emerge as we ascend to higher, more integrated levels? What properties, what talents, what realities will emerge out of that!? What can and will we become? Is there a limit? How do I find the answers to my questions?

Thus we end like we began—questions.
Our final words are, "Why? How? What?"
The words of explorers, adventurers divine.

So, of course, the final collapse of the final superposition becomes the initial conditions of the new superposition. That's nerd-talk for: The changes never end.
Thank God.

Epilogue: A Quantum Feast

It was an age of magic, when magic hung heavy in the air. The trees breathed their song to birds that told tales of power. Enchanted glens held secrets that the proper word would call forth gold to shimmer out of nothing. Magic was everywhere. And I was there. Immersed in the scintillating splinters of creation that I called real.

At first it was just a rustle in the distance. Trees shifting their feet and leaves breathing forth precious aire. Then the clip-clop as the horses wound their cobblestone way up out of old Vienna. Soon out of the myst one, then another, carriages appeared; the guests were coming! Doctors, Physicists, Mystics, Artists—Wizards all. Here to finish their book, and perhaps start another. To eat and drink and feast, and toast to adventures past won, and to adventures on the other side of tomorrow.

Inside the castle Sir Mortimer of the Cups, long appointed by the gods, scurried about in readiness. Although the grand feasts were many, he knew this one was special, and the harbinger of many, many more. "Something in the aire tonight," he hummed to himself. Perfection in the physical he deemed impossible, but nevertheless, to that he always strove.

The hall had been decked out in forest green, and the smell of juniper was intoxicating. Everyone had worked most diligently in preparation for the Great Feast. From the guests in

The question really is what is God doing to make a universe? That is the trick, and that's the trick that physicists try to find an answer to, and my interest has always been in magic as a way of approaching what I thought to be a pretty miraculous thing. Mainly, why are we here? What's going on? Even as a kid, I used to ask myself those questions. So what I've discovered, in answering the question of how this trick is done, this trick we call the universe is done, is to find that mind or consciousness cannot be disentangled from matter. That matter and mind are much more intimately linked than was ever thought before.

—Fred Alan Wolf, Ph.D.

their quiet studies and laboratories, probing into the secrets of life, to the poets and magicians ever searching for the muse, to those Quantum Cooks who were making kitchen things happen that had no precursor or successor. A blink-in of genius to delight the senses and spur great conversation.

For that, conversation, was the main dish in the main hall for the evening. And as carved in the stone archway of the great hall:

Great Minds talk about Ideas
Average minds talk about Events
Small minds talk about People.

But there would be no gossip about people this night. Everyone was looking forward to the interchange of ideas: new theories, new realizations, new emotions, and who knows!— maybe even a new addiction! Whichever way it went, tonight there would be conversation that was itself a force of nature:

Morphic Fields will be shaken
Cosmic Plenums unfolded
And Realities holographically crackling
Across Time and Space.

It seemed that everyone arrived at once.

At one moment Sir Mortimer was futzing over something in the corner, and in the next, everyone was everywhere. Fred Alan Wolf,[1] looking more and more like Dr. Quantum every day, was staring at a painting on the wall, pondering where it was a portal to. Mark Vicente strode into the grand hallway having covered the entire house. "You know, I could live here!"

Masaru Emoto and his beautiful wife had just come in from somewhere on the other side of the globe, but had found time in their relentless schedule to dine and laugh and create for a night. Suddenly, a shout went up.

---

[1] Okay, this is a story. And we're taking some liberties with the real people we've come to know and love. The later quotes are real. But we're nearly done, and it's time to celebrate! And enjoy "a tale told by a madman."

All eyes looked upward just in time to see Gordy, holding his and Betsy's baby Elorathea in his arms, launch himself (and her) on a mad slide down that long curving banister. It was quite a thrill for everyone, especially since Gordy was wearing his new leather kilt. He landed with a perfect two-point landing, on those feet that had conquered the firewalk, and so would never taste hot coals ever again.

And there was Betsy Chasse. Somehow the invisible hand and creator behind so so much of what went on, and ever a lover of costumes, she was dressed in some indescribable way, reminiscent of something no one could quite remember. Whether she continually changed through the night, or "just shifted," was a topic that outlived the gala ball.

As the traveling coats were gathered up and put away, the guests drifted into the salon. Comfortable couches were all about as the travelers sank in to pause and refresh themselves and prepare for the feasts to come. A delightful Mozart sonata drifted in from the music room, then ended.

Whereupon CHOAS erupted. Unbeknownst to the rest of the guests, Masaru had, like he had at the Water and Peace Festival in Tokyo, talked Ervin Laszlo into playing a little piece on the pianoforte. The "little piece" was a Bartok Romanian dance. A piece that forever bordered on chaos and mayhem, and at the moment when it seemed all hell would break loose, somehow out of it emerged a thread, a melody, that took the listener to the next precipice of insanity. Or as Dr. Laszlo would put it—a moment of bifurcation.

And then there was silence.

Many of the guests had not known that Ervin Laszlo's first career was as a concert pianist, and as a teen he had traveled the world playing with all the major symphonies. (It was even rumored that Bela Bartok's own piano somehow came to live in Dr. Laszlo's study.)

Wizards all, that dose of bifurcative madness blasted everyone out of their reveries and into the grand feasting hall. But this was no ordinary feasting hall, just as these were no

The wisdom of the mystics, it seems, has predicted for centuries what neurology now shows to be true. An absolute unitary being, self blends into other, mind and matter are one and the same.

—Andrew Newberg, M.D.

ordinary wizards.[2] Instead of a long table stretching across the room, it seemed that someone had taken that long table and wrapped it round into a donut. After a brief discussion it was decided that "Magicians of the Donut" wouldn't really work, and everyone opted for something a bit Arthurian: "Wizards of the Round Table."

As the gathering sat down, the three filmmakers glanced amongst themselves. Being the "hosts," it was their place to make the first toast of the evening. Who would be first stirred to eloquence? It didn't matter. Like an invisible hand reaching down and yanking him off his chair, Mark Vicente was up! He had something he had been wanting to say to the assembled guests for quite some time:

"I have been fortunate to have met truly great minds over the last few years. The knowledge and ideas I have learned from you and others have expanded my worldview enormously. Without all of you, I could not have done this. I give tribute to all it took for those who came before us who allow you to take their knowledge and build on it. The world is indeed a better place because of all of your efforts.

"Knowing what I know the amount of data I ignore—99%—I will do everything I can to learn new ideas with the most open-minded perspective I can muster. If there is so much that I don't know, I now understand that it would be foolish to insist on anything just because I want it to be a certain way. I want to see beyond my self-enforced blindness to look into the void of potential and ask the Great Questions:

"What am I in relation to the reality I see in front of me? How do I see what I do not know? How do I get out of my own way? If I am constructing the universe as I know it, what holds it together despite my emotional psychosis? What would it be like to not be invested in any one perspective but have the freedom to explore all perspectives?

"And, how would I develop the skill to stop presupposing?

---

[2] Is there such a thing as an "ordinary" wizard?

EPILOGUE: A QUANTUM FEAST

The answer for this to me is the making of a great mind.

"Thank you for helping me to rediscover the curiosity of a child and the tenacious critical analysis of a scientist."

And he paused. Mark, Betsy and Will looked at each other. It was a long, soulful look. It had been a trying, twisted road. Years ago, they had first encountered those sitting around the table. And asked questions and gotten answers. Some of which were not what they expected. They had been challenged; they had been befuddled; they had to go back and reexamine their prejudices and beliefs. [Will and Betsy rose to join Mark.] And those wonderful people sitting around the table had never criticized or denigrated the three filmmakers in their quest. It had always been about finding what is real, the truth, and enjoining in the discovery. What, what(!) a grand gathering.

Everybody stood up. It's like they already heard Mark's last words:

"To you, companions on the journey, and to knowledge!"

With a grand shout of "So Be It!" the goblets were joined, the bread was broken, and the feast began.

It was perfection itself. The food, the tales of power, the camaraderie. Sir Mortimer was everywhere, making sure the timing, the placement, the temperature of the seemingly never-ending array of dishes was just where it should be when it should be. It is rare indeed for an assembly such as this to have the time and leisure to enjoy a sumptuous meal and each other's company. Glancing over at Mark, Jeffrey Satinover said:

"I learned that if you want to really explore the most interesting questions in life, you are going to end up being wrong a lot, and there are going to be some people who are going to be right when you're wrong, and so you have to be willing to learn from that. Sometimes you're going to be right, and so you have to learn also how to be gracious when you're right if you want other people to listen to you.

"If you want to explore the world of what's really interesting, you have to just be accustomed to being confused and uncertain and allowing for the fact that there's so much that's mysterious, not using the exact nuance of the phrase 'being in the mystery' that Fred Alan Wolf uses, but sort of a more mundane use of the phrase that we all are so stupid really compared to how much there is out there that's mysterious that you have to have that kind of childlike sense of exploring the unknown."

And as the plates of steaming food were passed round the table, the stories kept rolling out . . .

From Candace Pert: "So my friend Deepak Chopra would tell this story how he was so excited about my work, and he went to India and he said to all the Rishis, 'It's unbelievable, this woman, it's wonderful, she's got the molecules, she's got the gels, she's got the receptors, she's got the peptides, it's unbelievable.' The Rishis are like 'what, what, what, what?' 'No, no, you don't understand. She's got the actual molecules of the emotion, there's the endorphins and the peptide, the hormones and the receptors, it's unbelievable.' They're all scratching their heads. He tries several more times. Finally, the oldest and the wisest Rishi sits up suddenly and he says, 'I think I get it. She thinks these molecules are real.'"

To which everyone laughs and guffaws . . . Enlightened nerd-humor . . . A rare treat indeed . . .

The meal was winding down. That night we could all say, as Ramtha often noted, "Tonight we dined like Kings and Queens." After-dinner drinks, double-triple cappuccinos and lattes, pots of PG Tips were left on the table, along with sweet delights and cheeses.

As the last cook left, the ensemble settled into their chairs in preparation for the main event of the evening: writing the last chapter of the UnUnWizard's HandBook Book. As they

I haven't studied the miracle of changing water into wine. It sounds like a good one.

—John Hagelin, Ph.D.

Every single cell in our body is spying on our thoughts.

—Joe Dispenza

ruminated on what their closing verse would be, a deep sense of relaxation, almost of laziness crept through the aire.

The center of the Round Table was open. A space of three or four long strides, so that the speaker, the one who "held the floor," could walk and gesticulate as they put forth their ideas. And ideas were the main course of the evening. And yet the diners sat, no one venturing into the center to begin the final chapter. A pipe was lit here and there, and an occasional cigar, and still they sat, lost in a moment of eternity.

But time was in fact slipping through the hourglass, and much there was to do. A book to write, another to unwrite, conversation and laughter, and, of course, toasts . . .

Will rose with one in mind: "When we set out to make the movie, one of our goals was to make the people sitting around this table Heroes. Heroes to the world. When I look around the table, I see everyone has given years of their lives to, as the alchemists of old said, the Great Work. And it has not been without risk and ridicule from colleagues and society at large.

"I do not think that because someone can, with an aluminum pole, redirect a ball over a fence 300 feet away, that they are a Hero. I do not think that spouting someone else's ideas in front of a camera makes one a Hero. Without the beings sitting here with us, and other explorers of the unseen, we would be in a stagnant, boring world, and not the magical one that we all love."

He paused for a moment . . . "Oh God, I'm getting maudlin. What was in that wine? I don't care! It's the truth. TO THE HEROES!"

Still unaccustomed as they were to such effusive toasts, the guests smiled and toasted. One of the lost arts in the modern age is the Art of Toasts: "Cheers" being shorthand for "I have nothing to celebrate or dream for." Intent rolled into the words rolled into the elixir soon rolled into the body and made intent real from the mind all the way through to matter. Minds linked together in that process make a very powerful magic indeed.

"I have one." It was Betsy. "To all the rest of the Heroes. To

> Usually in the supermarket, people say, "You were in that movie." And I say, "Yup, I was." And they say, "Wow."
>
> —Stuart Hameroff, M.D.

Is the observer going to be a meddling one who always meddles and tries to become the central part of the experience, or can the observer be a witness and let the experience unfold itself?

Some traditions are very good at it. The Hopi Indians apparently don't have a word for "I" or "we." They emphasize the verb, the happening. They would say raining, loving. See what is happening?

Ordinarily I make love to this person, right? But instead, if I say: Loving is taking place, then I'm only just witnessing. Loving is taking place; two people are involved. One me and one my significant other. And then what is happening is loving. There is no I, there is no it. It's just loving. See the beauty of that transition?

—Amit Goswami, Ph.D.

the people who take ideas and concepts and apply them to their lives. To those who thrive in the chaos and are living in the mystery and making known the unknown. I wish they were here with us!"

And with that everyone suddenly looked up from their goblets. Around the room. A thought quickly passed amongst them: They are here . . .

For if thought is real like the table before them is real, like the goblet in hand is real, and if thought moves outside of time and space, and if like-minded thoughts link up, entangle, that means that everyone whose mind is focused on that grand feasting hall is in the grand feasting hall. If in your mind you see the hall and everyone sitting therein, in that space—there you are also. The candles suddenly flicker as more guests pour in.

And time matters not. Years and years from that very moment, minds would whip back across time and join the Round Table discussion.

And who's to say who called who. Did the Round Table bring in beings for the discussion, or did the beings create the intent that the Round Table filled? Egos always want to be first, but in entanglement there is only happening. There's no difference; it's all the same impulse, producing both chickens and eggs.

Glancing around the room, JZ Knight, who sees things of a subtle nature, chuckled: "Oh my, it's getting pretty crowded in here. All these quantum mustard seeds. It's good there's lots of dimensions."

It was becoming a Grand, Grand Feast. And if the theories and the experiments were true, it was a feast that would go on and on. Ever evolving as more and more spirits popped in.

ith the thrill of a grand assembly, the gathered Magicians set about their task in earnest. Knowing that their verses would ripple out over time and space, they choose their words carefully, each vibration carefully intuned to the idea put forth. Since there were no Muggles about, the question of how to get inside the Round Table was no longer a priority, as they all simply blinked in and out. Except for Stuart Hameroff, looking like he just descended with the hordes from the steppes of Russia, who leapt up on the table, then down into the center in the arc of a perfect jump shot. He began:

"I think the next step is to try to explain how the quantum world can relate to our consciousness and to spirituality because I think that's the future in which science and particularly quantum physics and relativity come together with human consciousness, subconsciousness and spirituality. Whether a scientific explanation for spirituality is a good thing will depend on whom you ask. And I think if we explain everything, that will probably be bad, but I don't think there's much danger of that because all we're doing is peeling off layers of an onion that goes way, way down!"

As if to make the point, the moment he voiced the word "down," the floor seemed to suck Stuart away, and then he was back in his seat. Betsy had a twinkle in her eye:

"Okay, what is quantum cooking?"

A cluster of "probable, possible, superpositional" answers hung in the shared mind-space. Who would collapse it?

Betsy didn't give them time. Her costume shimmered, and suddenly she was a hawk, with a mask of hawk feathers and dazzling eyes. Hawk eyes looking for answers to her favorite question: "Why should I care abut quantum? Does it answer the Great Questions?"

Before anyone could respond, Dr. Wolf had already pounced: "Quantum physics is a starting question to the answer to all the Great Questions. It's a good place to begin. It's only the beginning. It's only the last hundred years that we even began to question that maybe we were barking up the wrong tree, asking the wrong questions, seeing the world as 'out there,' separate from the subjective experience of the 'in here' world.

> For where two or three are gathered together in my name, there am I in the midst of them.
>
> —Matthew 18:20

"Quantum physics begs that question. It says, wait a minute. There's a deeper connection going on here. I would say that what quantum physics is to the twentieth century, whatever is going to be the new bridging of science and spirituality will be to the twenty-first century."

He paused for a moment, then whirled around, addressing the entire room, seen and unseen. His eyes twinkled very brightly . . .

"The universe is random to a very, very large extent. And it's an important reason that the universe be random, to a very large extent. Randomness is the blessing, not the curse, of the universe. It allows something new to appear. What if everything were ordered and structured? We would all be robots, and we would not be able to think new thoughts.

"With randomness comes wackiness, and comes dance, and comes theater, and comes beauty—with it comes all the wonderful things of life. Because chance really makes life beautiful. Luck is a lady."

And everyone looked about them: "Yes, life is beautiful." They were all here to make it more beautiful, more wonderful. To take toads and remind them that they are Wizards, each and every one of them. To un-toad the toads. To reverse centuries of ignorance when humanity was kept in the dark. No more. A great shift had taken place, and secret knowledge was pouring forth, into books, seminars and the arts. The "Invisible College" was hidden no longer. The cave in Tibet had become the high-speed laptop. Information could no longer be kept from inquiring minds. It was magic of a different order for a different age.

Still Betsy wondered: How do we make all this real? Ever a Sicilian, Joe Dispenza drained his goblet and set it down before responding:

"So what do we have to lose, as human beings, to live as if thought is the supreme premise to infecting reality, and observing reality the way we want it? What do we have to lose? I'm a scientist, and I've been trained scientifically. However, by the same

means, it's all great dinner conversation, unless we have the abilitity to apply it in some way, shape or form. Now that becomes the true science. That becomes the true religion."

Meanwhile, Will had pulled out the original UnWizard's Handbook. A mighty tome, it had chapter after chapter extolling the virtues of shopping malls, sitcoms, gossiping about people, staying safe, having figured out that you have it all figured out, and that it's not paranoia if they're really out to get you. At the end of every chapter was a "Don't think about this . . ." section. A list of things not to think about to lead a not-as-bad-as-it-could-be life. Especially prominent was the "Wizards Who Thought They Could Fly" chapter. It featured all those who had wondered out of the box, their comfort zone, and did not make a million dollars. That fit in well with the ever-present depowering question: "If you're so hot, why aren't you rich?"

It seemed for a moment that a foul smell crept across the room. Will slammed the flatulent book shut: "The first UnWizard rule is: Convince people that they are NOT magicians."

Yes, rule number one. For in the truth of all matters, this limitation is the one that will surely stop all others. Aye! The self-imposed limitations are the hardest, well-nigh impossible to see, for the creator is in the creation, and by this all limitations are realized.

Will, well into his cups by then, suggested that the UnUnWizard's book could simply have on every page: "Look, you're a toad because you want to be a toad. Deal with it." Miceal suggested that might be a bit strident, however short and sweet it was, and that perhaps a more reasoned approach would suffice:

"The greatest problem that people have is not accepting their wretchedness, their poverty-stricken condition, their lack, their inability, their powerlessness. The greatest problem we as a human race have is accepting our own greatness. We

So do our thoughts matter? Indeed they do. They are the constructs of reality. What is a thought? Well, a thought is a frozen moment of a stream of consciousness that the brain processes and puts into a package called a neuron and then is added to by associative memory. So then you have a thought and you say, "Does this thought have meaning and power?" It does, because a thought is actually a structure in which reality is patterned. It is the architecture of reality actually. So when you create your day, you are composing in it thought, and as you observe the thought, it becomes the form in which reality itself molds. So the adventures of the day are really based upon your thinking.

—Ramtha

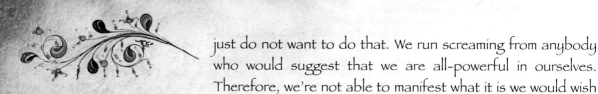

just do not want to do that. We run screaming from anybody who would suggest that we are all-powerful in ourselves. Therefore, we're not able to manifest what it is we would wish to have.

"If we could only accept who and what we are, and the real power that we have, then what we call the miraculous, which has shone forth in unfortunately all too few individuals in the past, that'll become commonplace. And we would learn the new science of manifestation, which is to realize that we have always, twenty-four hours a day, three hundred sixty-five days a year, been creating our own reality. There are no new powers to be learned. We already have them. What we need to change is the type of life that we are creating for ourselves."

JZ Knight continued the thought:

"When a person says, do I create my reality? The answer is, but you already are creating it. Who you are is made up of the reality that you are currently. To change that would be to change one of those concepts of people, places, things, times and events. The idea of who you are with, where you live, what you look like, what you wear, who you talk to today, what you're going to do tomorrow, you have all created this as fundamentally your reality, so everybody in your life is an aspect of yourself.

"We're so busy being it. It's like the fish in the ocean; somebody gives the idea to the fish that it's a novel idea to ask for a drink of water. So the fish asks for a drink of water, and everybody starts laughing. Because the fish is in the water.

"So it's sort of like saying, how do I create my own reality? Well, you are the reality; you're already creating it. We only see what we are when we step out of it, and we look back at who we've been."

If reality is my possibility of consciousness itself, then immediately comes the question of how can I change it? How can I make it better? How can I make it happier?

—Amit Goswami, Ph.D.

t that conversations erupted around the table. It seemed that everyone was talking at once, for all the conversations were interrelated. Spirit related to matter related to consciousness related to creating related to intending. Emotions, neuronets, old para-digms, back to consciousness, observer, choices and change. And reality—the intermediary concept between everything—was itself defined by everything.

How can one define a word except by using words? Is it possible to understand a concept without other concepts? If that is true, how do we ever truly know? At this turn of the conversation, Betsy had her final toast:

"You know for me, it was all just philosophy, which I love, but only when I coupled that with practical experience did it light up my life."

Bingo!

The fabulous interchange of ideas that had lured everyone from the four corners had arrived. The world of human thought had been all about boundaries for ages: "What is mine; what is yours? This discipline does not talk about that. Don't use the word entanglement—you don't even know what it means." But in the end, the narrow, narrow avenues of scientific, philosophical pursuit had run headlong into a box canyon. At the end of particles, you only found more particles. At the end of disease, you never found health, only disease.

The focus that had allowed Western civilization all its marvelous advances had cut it off from the world of magic. But as the night stretched past midnight, a sense unfolded that in some corner somewhere, something had been turned. The paths were coming back together. The grandeur of the dream to find the simplest explanation for the most widely observed phenomena was like an elixir—and in that moment an activity long thought gone, relegated to a bygone age, was rediscovered.

Conversation, inquiry among friends, glorious theories and ugly facts—this was a feast worth having. And they knew, everyone knew, that the ripples were going out out out into the world. Just as likely, it was the ripples coming in in in from the world that drew them together in the first place. The holographic universe repeated itself in miniature everywhere.

> A sign of boredom is a sign to change. And what is it when we change our mind? We change where our observer sits in our brain. That begins to fire new neurons, which the rabbit hole is like a wormhole that loops us into a new neuronet in the brain.
>
> —JZ Knight

And to a degree, these Magicians made a breakthrough so other Magicians everywhere would have that possibility. As goblets were raised in celebration of the triumph of the human spirit in one home, it would surely as the sun would rise, reverberate in homes and dwellings, inns and taverns everywhere. Such is the nature of reality. Such is the universe we find ourselves in.

Long since had the guests pushed away from the Round Table and gone to more comfortable surroundings. Some removed to the salon, to sit by the fire and tell stories. For all pilgrims on the way back home have harrowing tales of fright and power, and it was a joy for all to sit on the other side of them and laugh.

At times the smoke seemed to gather strangely, and other guests were seen in the haze. Certainly no one of this group seemed to mind. The word went out that it was just "old friends coming by for a peek." The candles would flutter, and a wind would whip through, and the flames were taken to the brink of extinction, when suddenly the wind was gone. "A cheap parlor trick," someone noted, whereupon his cigar ash fell into his lap. "Touché!" And no one laughed harder than the one with the ash. Such is the humor of Magicians.

One by one the gathering quietly dispersed. A few meandered upstairs where they had rooms for the night. Occasionally a car could be heard pulling up, a door opened and shut, and then silence. There was rumor of a Lamborghini.

One of the scientists decided to "stroll the old city until dawn," after which he greeted the new day from atop the Eiffel Tower.

The last to leave was Dr. Wolf in his splendid carriage pulled by four horses.

It was afterward said that when the guests looked from their second-story bedrooms that the carriage looked more

> The only way I will ever be great to myself is not what I do to my body, but what I do to my mind.
>
> —Ramtha

and more pumpkin-like as it went down the avenue. Smaller also with every step of the horses. There were initials carved into the back of the carriage: "F.A.W." But an instant later, the initials seemed to be a grinning face, with light pouring out of the features. Someone heard a chuckle, then a pop, and they were gone.

And so it was that the first rewriting of the UnUnWizards HandBookBook came to an end. Ere the sun had fully risen, the guests had scattered to the four corners once again. Like the bards of olde, there was always the next town, the next adventure, the next "unknown," crumpling under the impulse of . . . of . . .

"This life is but a page in an enormous book," says Ramtha, "in which we will always be who we are. But always with the inherent needling of ambitious pursuit. A pursuit that takes us from the boring tedium of self-reflection, of self-hate, to self-creation of new dreams, models of thought that do not bridge the insane or the redemptiveness of failure, but new models of thought that we participate in with the zeal of ambitious energy."

Yes, that's it, ambitious pursuit. Making known . . .

And as I climbed the long staircase, blowing out candles as I went, I thought to myself, and to everyone else out there who was still listening: "Yes, magic is everywhere. Especially tonight. And every night. Ahh, what a night it's been . . ."

And with that, we bid you all
A most fond fond
Adieu.

The End

The Making of

What tHē BLEEP Dō wΣ (k)πow!?™

from keynote speech by Will Arntz
at the Prophets Conference in Santa Monica 2005

People often ask the three of us how this movie ever happened. I think it's one of those things my colleagues and I must have arranged before we ever got here—you know, there you are on the Plane of Bliss, you're shopping around for new bodies, and an angel comes up and says, "Hey, you guys, want to make a movie?"

"Yeah, that sounds like . . . is it going to be hard?"

"No, no, no, it'll be great. You'll go down there, and wonderful things will happen. You'll have a great success and hold conferences. . . ."

There was a point a few years back when I had sold my first software company and retired. I soon found myself bored—well, actually more pissed than bored, because I had sold the goose that laid the golden eggs. I wrote the software, sold it and did okay, but the people who bought it did better—they became billionaires.

So I decided to do a second company, and I had the money to finance it. I was in New York City working on it, and there was a key moment that, in retrospect, helps all this make sense. A friend of mine had done a play Off Broadway, and I had helped a little bit to finance it. It was about Buddhism and that kind of good stuff, and it was the neatest feeling to help finance something like that.

One day I was walking toward the World Trade Center, of all places, thinking about my new project. It was just at the time that Ted Turner had pledged to give a billion dollars to the UN. And I thought, "Well, that's great. But is it really going to accomplish anything? He's just pouring money into the existing power structures, so we know what's going to happen: the same old, same old. Someone needs to create works that will actually change consciousness. Somebody needs to do that, and someone needs to finance those projects."

And suddenly the lightning bolt hits: "I can do that. I can develop more software and make more money and do that." And I thought, "If not me, then who?" It seemed strange that it happened to be me, but I figured, "Okay, let's do it! I'll start a $6 million software company and use some of the money for good projects."

Now, I had begun going to Ramtha's School for Enlightenment a year or two earlier, and there's a lot of focus there on creating your reality, holding your desires and intentions in your mind, and then manifesting them. So I thought, "You know, I'm going to apply these principles. Instead of just doing a $6 million dollar company, I'm going to do a $30 million company, and half the money I make I'm going to donate to works of the spirit."

So in that moment, walking toward the World Trade Center, I kinda cut a deal with the spirit. Okay, you're going to get half, but I may need some convincing on the side, too, to make this thing happen.

My first teacher used to have a sort of

bidding war with the universe. The universe does something good for you and you go, "Oh, you think that's good? Ha! Here's what I'm going to do for *you*; I'm going to give back twice!"

And the universe goes, "Fine, but now I'm going to give back to *you* twice."

So you go, "I'm still not impressed . . ."

This is what's going on as I'm walking toward the World Trade Center. Inside I'm thinking, "Rock 'n' roll, we gotta deal?" And I do the secret inner handshake.

So I do the second company, and make the money, and retire. Now, when I retired I didn't think I would be the one to actually do the work. The goal was to provide the money, and other people would come and say, "I've got this great project," and I'd say, "Yeah, go ahead and do that" and write a check, and then they'd send videos of what they were doing and I'd watch them. Well, it didn't quite work out that way.

At some point, when I was up at the Ramtha School, I got introduced to Mark, who was editing a movie. I had done movies—not professionally, just in graduate school. I started watching Mark over his shoulder and thought, "Wow, that's kinda fun. Maybe I could do something like that."

One night at the school, Ramtha was walking around on the stage saying, "In this school we talk about quantum physics, molecular biology . . ." and he lists the sci-

THE MAKING OF *WHAT THE BLEEP DO WE KNOW!?*

ences we were studying and says, "Somebody should write a book about it." I'm sitting there, and like a wise ass I think, "Write a book? Someone should make a movie."

It was one of those moments where you realize, "Oh shit, I just volunteered for something."

A month later it's still going around in my brain, and I say, "Okay, I'm bored anyway. I guess that someone happens to be me."

So I talked to JZ Knight and started working on this thing. The original idea for the movie was to take a lot of the teaching footage of Ramtha that had been filmed over the years, cut that together with some animation, add some interviews and maybe do something that, if we were lucky, would be a PBS-type movie. If we were lucky. And it was going to cost maybe $100,000 to $125,000.

I start looking at footage and putting this together. And I start writing little skits to illustrate the weird, wacky quantum world. One of them actually made it into the movie—the bit with Amanda on the basketball court, with the multiple basketballs bouncing. That's about the only thing that lasted from the first script.

As I start getting more deeply into this, I think, "Maybe I want something a little bigger. If I do, I need help." So I turn to Mark and I say, "*Help!*"

Pretty soon the two of us are working together on scripts, and we're all enthused.

We're working and working and working, writing funny little things. At this point Mark realizes that we need help. I was so far over my head that I didn't even know it. He says, "We need a real producer."

I say, "Really?"

"We gotta talk to Betsy."

"Okay. What's she doing?"

"She's making gourmet dog treats."

So I call up Betsy. Meanwhile, Mark has given her the script. She reads it and thinks, "Wow. This is amazing; it answers a lot of questions." When I get on the phone with her I ask, "Why do I need you as a producer? I can hire a line producer."

She goes, "Yeah, you can hire a line producer. But when the grip truck doesn't show up because the union doesn't do that, and the drivers are on strike because they didn't do the eight-hour turnaround and blah blah blah," and I go, "You're right. Maybe I do need help."

As I'm getting ready to fly down to meet her in L.A., I get a phone call and a friend asks, "What are you doing?" I say, "Getting ready to go to the airport." He says, "You're not going anywhere today." I say, "Why?" and he says, "Have you turned on the TV?" I turn it on in time to see the second World Trade Tower go down.

I didn't know what to do. I wanted to hire Betsy, but I didn't want to hire anyone without actually meeting them because, you know, she could be wacko. So we talk some more on the phone, and while I'm trying to

decide what to do, she tells me a really sweet story. She has this thing about stray animals; she takes them in and cares for them. One day, a bird flies into her bathroom. The cats are going after it, so she shuts the door and starts talking to the little bird; it flies into her hand, and she walks outside and lets it go.

That was the moment I decided to hire her. Now, of course, as soon as I hired her, she became little Miss Truck Driver: "Well, that sucks. I can't believe those idiots did that. Ranh ranh ranh. . . ."

I said, "Betsy, what happened to you?" And she said, "Ah, get used to it!"

So now there are three of us working on the script. But by now something has shifted. We all got the crazy idea that we wanted this to be theatrical. PBS was not good enough. Discovery Channel—unh-unh. We wanted to be in theaters around the country.

We had the sense that there were millions of people in the world hungry for this information. We could feel it. We could feel these people wanting a new worldview, wanting to see things differently. "Whatever we've been taught, whatever we're doing, it ain't working." We felt this really strong pull, which is one reason we didn't pull back when we were making the movie and everyone in the industry said, "You guys are crazy. You can't make a movie about this stuff interesting. And if you could, no one would come." So the big joke is, all the hundreds of thousands of people who've seen the film—probably millions by now—don't exist. Just ask Hollywood.

Now, it would be great to be able to say that our intent was clear, we had a strong focus, and it was a manifestation of our inner knowingness that this was going to happen. Ha! It's more like Spanky and Our Gang. Remember Spanky and Our Gang?

"Hey everyone, whaddya say we make a movie that changes the world??"

"That's great, Spanky. How we gonna do that?"

"I know! We'll get a camera!"

Mark and I had been developing this script for quite a while. Pretty soon after Betsy showed up she says, "You know, guys, what you're making here is a Discovery Channel movie. You've got a host." It was true; at that point we had a host, walking around talking to the camera, saying things like: "If you think quantum physics is strange, look at *this* phenomenon." That's the way it was written.

So Betsy starts her campaign to get rid of the host. Mark and I ignored it and ignored it and ignored it . . . and kept hoping. . . . We kinda knew she was right, but we couldn't quite figure out what to do at that point.

Chaos was nipping at our heels the whole time. (And sometimes it got a little higher!) One day one of us said, "We don't know what we're going to do. We have no idea how we're going to do it. So, let's go interview people. Maybe the people we interview will help us. Maybe we'll learn stuff, and that will give us some direction, because we obviously have none."

So we started calling people, saying we wanted to interview them. We set up meetings and started interviewing. For me, one of the big moments happened right near the beginning, when we interviewed Ramtha. I got the nod to do that. The way it turned out, my interview with Ramtha lasted ten seconds. His interview with me lasted three hours. He said, "You want me to sit here like a mannequin and answer all your questions, so you can have a great movie? Guess again! I just have to make a better you, and you'll do all the work."

As he said later, I needed humbling. Because the way it was, I was financing the movie and paying all the bills, so I figured I got to call the shots. He said, "Making this movie and everything you're doing is just to

set up a certain emotion. To derive a specific emotional response. But the thing is, that emotion that you're going after is from the universe you know. You're controlling the outcome of everything just so you can have a certain emotion. And guess what? It's an old and lousy emotion that you've had for a long time."

I went, "Yeah, I guess."

And he said, "No, no, no! You don't realize the severity of the problem. You've been the same asshole for thousands of lives."

So there it is. I'm sitting there in the cross hairs. Mark and Betsy are going, "Whew, I'm glad somebody finally told him." The cameras are rolling. The crew is laughing. And that was just the lob for the serve.

"You want genius," Ramtha said, "but genius always comes from outside of the known, from outside of control. If you had made the movie you wanted to make, we would have a piece of shit that no one would want to watch!"

The thing about it was, it was all true. I knew it was true. He had me. He had all of us. Because we all did want something great, and we had reached the point where we were willing to get rid of the ego as best as we could to make it.

Teachers love it when their students really want something. "I can really use the old thumbscrews now . . . I can really put the pressure on." Because it's only by going through the chaos and into the unknown, and grappling with the unknown, that you'll have genius. I sometimes have an image of Ramtha sitting, wherever Ramtha sits, and he has this little button in front of him labeled "Chaos" and he just can't help himself. "Ah, they think they're doing this. . . . Beep." And it would all blow up.

So that was the big turning point. Afterward I was still shell-shocked when the three of us get together. Betsy goes, "Yeah, we've got to blow everything up! . . . Let's get rid of Amanda."

Mark and I look at each other and say, "No, I think we need Amanda."

So she says, "Okay, let's get rid of the host!" She'd only been trying to do that for three months.

"Yeah," we said, "let's get rid of the host." So in that moment we threw out six months or a year of work. And we set out to interview people.

Because I'd had some science background and could do the secret nerd handshake, I would lead off and ask the more technical questions. Then about the time they were starting to get a little hazy-eyed and had had enough of me, Betsy would come on. And she would go, "You know, you guys have been talking about all this stuff, and maybe it means a lot to you eggheads, but what does it mean to me? I mean, quantum superposition. So what?"

She didn't do it in an arrogant way. It was more like, "What does this mean to me? How does this affect my life?" And that

THE MAKING OF *WHAT THE BLEEP DO WE KNOW!?*

would bring a whole different response out of the person, and suddenly everything would become electric again.

Then Mark would come in. After sitting on the side listening and observing, he would sometimes come up with a real showstopper.

We did all the interviews we wanted—and stepped back into chaos world. Okay, we have no script, and sixty hours of raw footage. What are we going to do?

Meanwhile the budget is escalating. By this time it's over a million dollars. I tell myself, "Okay, okay, it is what it is."

The plan—and the challenge—was, could we intercut pieces of the interviews so that it all made sense? So that it would seem like they were having a dialogue? We didn't want to get our own faces on the camera. We wanted to remove ourselves from the whole thing. Eventually, the intercutting kinda worked. I edited it down to about 2.5 hours, arranged more or less in the order you see it in the film.

And then came the dark night of the soul.

Or, the dark night of the *BLEEP,* because now we had to write the script. And we didn't know what to do, because we had shot the host in the head six months before. So how do you get all this stuff out and have a story line, without someone kind of leading it along? We went through dozens of ideas. We went through the idea *du jour.*

Sometimes one of us would have The Revelation in the shower in the morning and come running in: "Hey, I've got it figured out." Once Betsy came in and said, "Okay, Amanda is a woman who types transcripts for movies. And all this footage comes in, and she starts interacting with it." It was a pretty good idea, but both Mark and I went, *ennhhhhh. . . .* Then maybe I would get an idea and the buzzers would go off. It was a difficult time. We had to make ourselves a *cheeseometer . . .* because some of the ideas we came up with were just so hokey and cheesy. We actually had one in the editing room, too.

After six months of brilliant ideas, we still don't have a script, or even a title. We're taping 5x7 cards on the window, trying to put this whole thing together for the hundredth time.

Betsy would often say, "Ah, what the fuck do we know?" or, "What the fuck do I know?" Well, one day Mark and I picked up on that clever phrase. Because we realized, what the fuck *did* we know? We can't seem to figure this thing out. We thought we could figure it out and we can't . . . and then someone, I don't remember who, says, "Maybe we should call the movie that, ha ha ha ha ha." So we write it down and put it up there and figure at some point we'd come to our senses. But guess what? We didn't. The name stuck.

One reason we stuck with it is that we

didn't want to come off as saying, "This is the way. This is the way the universe works. We have it figured out. You know, it's our way or the hell way. Either you agree with us or you fry."

After months and months and months, little things start to click. At one point Mark says, "I think Amanda is a photographer." Okay. That works. She's a photographer.

Finally, we have the script. And I think— because I'm still Spanky, right?— "Wow. Gee, we have the script; now it's gonna be easy. We're just gonna film what's in the script and it's gonna be great." Of course, it didn't work that way.

When we started sending the script out to actresses to play Amanda, pretty much across the board, people respectfully declined. Why? It's a weird movie that you can't make interesting, and no one's gonna come to it. Everyone knows that! So what are we gonna do?

Then our casting director did what's called a "breakdown service"—you put the script down for people to look at—and Marlee's production partner, Jack Jason, stumbled across it. We had written it so there wasn't a whole lot of dialogue in that part. We already had an hour and a half of talking heads, and we didn't want much dialogue. He said, "This is perfect for Marlee because she communicates so much nonverbally."

My first reaction was, "What, are you kidding? No, no, no, no, no, no, no." But Mark and Betsy both said, "Well, this could work."

People always want to know how it was for the three of us to work together. Think of three cats, fighting and scratching all the time until they get tired and flop over. Then the one who has some energy gets up and runs with the ball for a while. At different times, each one of us would take charge, and this was a time when Mark and Betsy took the lead. We had a meeting with Marlee, and we all felt, "Wow, it's gonna work."

THE MAKING OF *WHAT THE BLEEP DO WE KNOW!?*

We had never conceived of Marlee being that part, but it felt right. We got the idea that maybe the *film* was making the choice at this point. When you get involved in a creative project, it takes on a life of its own, and at a certain point the artist just has to shut up and get out of the way. So we thought, "Let's just shut up and get out of the way. Marlee showed up. It's an excellent idea. I wish it was ours, but let's go with it."

We filmed the whole thing in Portland. It's a wonderful city. It was an adventure. All along the way I would think, "Wow, now that we've got *that* done, it's gonna be easy." And then the old chaos button would get pushed again.

When we wrote the script we wrote it in a somewhat modular fashion, so different parts could go in different places. We thought that would be really clever, and that it would give us a lot of options in the editing room. It did—way too many! So once again we were back in chaosland. We didn't know how to put the thing together. There were a hundred different ways to wire everything—ten of them were good, ninety of them weren't. We spent a year and a half in postproduction, editing, bringing in the music, doing the animation, working it, working it, working it. And while we were doing this, we were also doing a lot of test screenings.

Because it was such an unusual form, we had to make sure that we weren't just crazy. So we would bring in an audience. Doing these test screenings was a way for us to get a sort of "beginner's mind," like they say in Zen. We were used to all the nuances, but we found that when we sat with a group of people who had never seen it, we could sort of hitchhike along on their awareness and see it as if for the first time. So we did a lot of test screenings and saw what worked. And we kept cutting and cutting. I think we did just about twenty major versions of the movie. And we finally got the thing done, three years after starting it.

It was very exciting. It was finished. People seemed to like it, so we arranged for potential distributors to come see the movie. We had created a marketing plan saying that an audience for the film really did exist even though Hollywood doesn't believe it. When they came to see the film, to a one they said, "Hmmm, interesting movie, no audience, good-bye."

At that point I fell back on my software experience. When you have a new technology, you have to do a "proof of concept." We had to do a proof of concept to show that an audience for this movie actually existed. Betsy went to our local theater in Yelm, Washington, and said, "Can you give us a screening?"

"Oh, I don't think so."

"Oh, please, will you give us a screen?"

"No, not really." Back and forth and back and forth. Finally, just to get rid of us, the person says, "Okay, we'll give you a screen. But you know, you'll probably only last one week."

It played for seven weeks. The media still ignored us. And the guy at the theater— after about four weeks, he said, "Yeah, I knew this was going to be a big success." We were so happy, we just said, "Right, thanks bro."

In Portland it ended up playing there for eighteen weeks. So now we're a success in two movie theaters. We did everything backward from Hollywood; it's the inverse of the Big Opening.

The other inverse, and the reason Hollywood people thought no one would come to see the film, was the underlying assumption of the movie: that our audience was smart. Hollywood's basic assumption, of course, is that you guys are stupid, and we have to spell everything out for you. It's gonna be everything you've seen before, we're just going to turn up the volume a little bit. We said, "No, people are smart. They *like* using their brains." In fact, one of our agendas in this whole enterprise is to revive thinking, which has become a lost art in this culture. Mark likes to say that he wants to make thinking sexy again.

So we started our march—a theater here and a theater there. At first we had to work really hard to find our audiences. We would get hooked up through Unity churches and Science of Mind folks. We'd go to yoga studios. I think we opened up next in Tempe, Arizona, where we're still playing. And this is a one-screen movie theater. It's not like we're in the corner of the quadraplex.

We kept doing it that way, one by one, until eventually we found a distributor. Samuel Goldwyn Roadside Attractions had released the film *Super Size Me,* and we were in a lot of the same theaters. They kept seeing this weird movie with the strange title, which would do better than their movie. We had no PR, and *Super Size Me* had a lot. So they started watching us, like, what the BLEEP is going on?

Eventually we had a meeting with them in Santa Monica. We were about to open on the Promenade there; we were moving over from the Beverly where we had been playing for weeks on end. They said, "Did you take out a big ad in the *L.A. Times*?"

I said, "No, we didn't do any ad in the *L.A. Times*."

They said, "That'll be interesting. We've been in this business for years, and you have to do an ad."

I said, "Yeah, okay, gee, that's why I want to work with you guys, because you know all this stuff."

So we walk around the corner to the theater, and there's this big sign that says, "*What the BLEEP* sold out."

We go in, sit down and watch the whole movie. There's a Q&A afterward. They listen to the discussions that are going on, and the questions, and the number-one guy says, "Wow. I've never seen anything like this. Yeah, let's do it." So we started doing our distribution.

Now, in my perennial role as Spanky, I figured people are either going to love this movie, or they're just going to ignore it. None of us had ever been in the public eye before; we'd never experienced the media, and we didn't really know what to expect.

Then the hate mail started coming in. Death threats were coming in. Oh my God, it was strange. At one point Betsy and I stopped

reading the guestbook on our Web site because you'd get three or four comments, "The movie changed my life"; "It's so wonderful," and then you'd read the next one and it's, "You Nazis. The world would be better off without you on the planet." I wasn't used to it.

It got so bad that I ended up calling JZ Knight, who channels for Ramtha. She has gotten her share of hate mail over the years, and she helped me a lot. You have to unhook from everyone's reaction to what you're doing, she said, whether they love you or they hate you. And some people *are* going to love you, and some are going to hate you. In the end, when you're sitting around at night drinking your tea and staring into the fire, you have to be happy with what you've done. You have to make an honest appraisal of who you are. And if what you've done you find admirable, let that in. If you fall apart every time someone doesn't like what you do, you're never going to leave the house.

It was a big learning experience for all of us.

Almost from the beginning, we started reading that the movie had been financed by the Ramtha School, and it was a recruitment video for them. It was so bizarre for me to read this because I knew they didn't finance it—I did. I wish they had—I'd have a lot more money left!

Then in the *San Francisco Chronicle* this woman wrote a review in which she referred to the people we interviewed as "so-called experts." Now here's a journalist—if you can call her that—who has decided that the people we interviewed were "so-called experts." I mean, if you don't like the movie, if you think it's silly, if you think that *artistically,* fine. But to pick on the scientists who have spent years of their lives coming up with these discoveries? In our minds, they're the real heroes. Not the guys who hit balls with baseball bats. These people who have done the research, they're the heroes.

When I went to bed that night, a Bob Marley song started going through my head. It's called the "Redemption Song." There's a line that goes, "How long will they kill our prophets while we stand aside and look?" And that just kept going through my head.

The way this society is, the press can just kill our prophets by making up stuff. And we all just go, "Oh yeah, it's the media, what do you expect? Oh well." We just roll over and say "oh well" *while we stand aside and look.*

I woke up the next morning breathing fire, and I wrote an e-mail reply to the person who reviewed it. And I said to the people on my e-mail list, if you want to reply, here's her e-mail.

Over 500 people wrote the *San Francisco Chronicle,* and that's an amazing number. And what did the *Chronicle* do? Nothing. They didn't print one thing. And we realized, "Oh my God, this is the media."

It's been so interesting for the three of us. Essentially, what we've done is stood up and put our stake in the ground and said, "This is what we think is important." And learned a little about how to deal with both the negativity coming at us and the positivity.

And that's pretty much the BLEEPing story about how the BLEEP we got here. And we have no BLEEPing idea where we're going next. But we're all very happy to be on this wild ride and thank you all for coming along.

Further Antics from the Folks at

What tHē ßLēēP Dө wΣ (k)πow!?™

   We have been besieged by requests to make the entire interviews available, either in printed form or a DVD of just the interviews. It seems that people want more information (which is why we wrote this book).

   But we decided to take it one step further. At the Santa Monica conference, we interviewed almost everyone who had been in the film again. Then we tracked down Lynne McTaggart and Dean Radin and interviewed them, altogether adding thirty hours of interviews to our existing sixty. While we were at it, there were about ten minutes of animation that we wanted to have in the first version, but ran out of time, so we went into production on that.

   Then it was back in the edit room. Ninety-five percent of the interviews from the first version were taken out and replaced by other footage. We added another hour of interviews and the animation, restructured the dramatic part and called it *The Rabbit Hole Version.*

   But then, why not take it one step one step further further? Utilizing DVD technology and the random number generators (REGs!) built into the players, we produced a six-hour version where the viewer can set "how far down the Rabbit Hole" they want to go. And if they want to go quantum, they can turn on the random feature, so that at run-time the DVD player "rolls the dice" to see what interview to play next. (God may not play dice with the universe, but we can do it with the DVD.) So every time you view it, it's different. (And if consciousness affects REGs, what does *that* mean?)

Contributors

David Albert, Ph.D., professor and director of Philosophical Foundations of Physics, Columbia University, specializes in Philosophical Problems of Quantum Mechanics, Philosophy of Space and Time and Philosophy of Science. Professor Albert is the author of *Quantum Mechanics and Experience* and *Time and Chance,* and has published many articles on quantum mechanics, mostly in the *Physical Review.*

Joe Dispenza, D.C., studied biochemistry at Rutgers University in New Brunswick, N.J. He received his Doctor of Chiropractic Degree at Life University in Atlanta, Georgia, graduating magna cum laude. Dr. Dispenza's postgraduate training and continuing education has been in neurology, neurophysiology and brain function. He is the recipient of a Clinical Proficiency Citation for clinical excellence in doctor-patient relationships from Life University and a member of the International Chiropractic Honor Society.

Often remembered for his remarks on creating his day in *What the BLEEP,* Dr. Joe is a student of Ramtha's School of Enlightenment, a contemporary school of ancient wisdom located in the United States, where he has learned to create his day and personally experienced how the brain, consciousness and intent work together to create reality in many various forms, whether it be a day, an event, an object or a future. His new DVD series, *Your Immortal Brain,* looks at the ways in which the human brain can be used to create reality through the mastery of thought.

The research of **Masaru Emoto, Ph.D.** (*www.masaru-emoto.net*) and his stunning photographs of water crystals have become the source of great widespread interest. Dr. Emoto's books include *The Message of Water, Vols. 1, 2* and *3*, the *New York Times* bestseller *The Hidden Messages in Water* and *The True Power of Water.*

Amit Goswami, Ph. D., is professor emeritus in the physics department of the University of Oregon, Eugene, where he has served since 1968. He is a pioneer of the new paradigm of science called Science Within Consciousness. Goswami is the author of the highly successful textbook *Quantum Mechanics.* His two-volume textbook for nonscientists, *The Physicist's View of Nature,* traces the decline and rediscovery of the concept of God within science.

Goswami has also written eight popular books based on his research on quantum physics and consciousness, including *The Self-Aware Universe, The Visionary Window, Physics of the Soul, Quantum Creativity* and *The Quantum Doctor.*

In his private life, Goswami is a practitioner of spirituality and transformation. He calls himself a quantum activist.

John Hagelin, Ph.D., is a world-renowned quantum physicist, educator, author, and science and public policy expert. Dr. Hagelin has conducted pioneering research at CERN (the European Center for Particle Physics) and SLAC (the Stanford Linear Accelerator Center) and is responsible for the development of a highly successful grand unified field theory based on the superstring. As director of the Institute of Science, Technology and Public Policy, a progressive policy think tank, Dr. Hagelin has successfully headed a nationwide effort to identify, scientifically verify and promote cost-effective solutions to critical social problems in the fields of crime, health care, education, economy, energy and the environment. In addition, Dr. Hagelin has spent much of the past quarter-century leading a scientific investigation into the foundations of human consciousness. In his book, *Manual for a Perfect Government,* Dr. Hagelin shows how, through educational programs that develop human consciousness, and through policies and programs that effectively harness the laws of nature, it is possible to solve acute social problems and enhance governmental effectiveness. In recognition of his outstanding achievements, Dr. Hagelin was named winner of the prestigious Kilby Award, which recognizes scientists who have made "major contributions to society through their applied research in the fields of science and technology." The award recognized Dr. Hagelin as "a scientist in the tradition of Einstein, Jeans, Bohr and Eddington."

Stuart Hameroff, M.D., (*www.quantumconsciousness.org*) is a professor in the departments of Anesthesiology and Psychology, and director of the Center for Consciousness Studies at the University of Arizona in Tucson, Arizona. Dr. Hameroff spends most of his professional time as an attending anesthesiologist at the University of Arizona Medical Center, caring for surgical patients and teaching residents and medical students in a busy trauma hospital. His research interests have always focused on how pinkish-gray meat—our brains—produce thoughts, feelings and emotions. Is the brain merely a computer? Or is something deeper going on? Might our brains be connecting us to the fine structure of the universe? What IS the fine structure of the universe?

Dr. Hameroff has published hundreds of scientific articles, including three co-authored with Roger Penrose, and five books including: *Toward a Science of Consciousness I-III* (MIT Press) and *Ultimate Computing: Biomolecular Consciousness and Nanotechnology* (Elsevier-North Holland). As director of the Center for Consciousness Studies (*www.consciousness.arizona.edu*), Hameroff organizes the conference series "Toward a Science of Consciousness."

Ervin Laszlo, Ph.D., is the author or editor of seventy-four books translated into twenty languages, and has over four hundred articles and research papers and six volumes of piano recordings to his credit. He serves as editor of the monthly *World Futures: The Journal of General Evolution* and of its associated *General Evolution Studies* book series. Laszlo is generally recognized as the founder of systems philosophy and general evolution theory, serving as founder-director of the General Evolution Research Group and as past president of the International Society for the Systems Sciences. He is the recipient of the highest degree in philosophy and human sciences from the Sorbonne, the University of Paris, as well as of the coveted Artist Diploma of the Franz Liszt Academy of Budapest. His numerous prizes and awards include four honorary doctorates. He is an advisor to the UNESCO Director General and ambassador of the International Delphic Council. He was awarded the 2004 Goi Peace Prize in Japan, and he was nominated for the Nobel Peace Prize in 2004 and renominated in 2005.

Miceal Ledwith, L.Ph., L.D., D.D., a native of County Wexford, Ireland, was ordained a Catholic priest in 1967 after studies in arts, philosophy and theology. After completing doctoral studies, he was appointed a lecturer in theology at the Pontifical University, Maynooth, in 1971, and became a professor there in 1976. He became dean of the faculty of theology in 1979, vice-president of the college in 1980 and served a ten-year term as president starting in 1985.

From 1980–1997, Dr. Ledwith served three terms as a member of the International Theological Commission, a small group of theologians who advise on theological matters referred to them by the Pope or the Congregation for the Doctrine of the Faith. He has been a student of the teachings of Ramtha's School of Ancient Wisdom for many years. As an introduction to some of the main themes of his forthcoming book, he is currently producing a DVD series entitled *Deep Deceptions*.

Lynne McTaggart is the award-winning author of the bestselling book *The Field (www.livingthefield.com)*, which grew out of her quest to see if new scientific theories could explain homeopathy and spiritual healing. She's best known as founder and editor (with her husband Bryan Hubbard) of leading newsletters and books on alternative health and spirituality, including the international newsletter *What Doctors Don't Tell You*. She's also written a book called *What Doctors Don't Tell You*. She resides in London and New York with her husband and their two daughters. Her company holds regular *Living The Field* workshops and conferences, where the world's leading scientists and quantum physicists speak and enlighten, side by side with healers and remote viewers. Lynne is now at work on a follow-up to *The Field*.

Daniel Monti, M.D., received his Doctor of Medicine, summa cum laude, from the State University of New York at Buffalo School of Medicine in 1992. His postdoctoral work was in the Research Scholars Program, Department of Psychiatry and Human Behavior, at Jefferson Medical College, Philadelphia, Pennsylvania. He is currently Executive and Medical Director, The Myrna Brind Center for Integrative Medicine, Thomas Jefferson University.

Dr. Monti's research has covered a broad range of topics and include grants in: Mindfulness-Based Art Therapy for Cancer Patients, Alternative Therapies in Disease, and Muscle Test Responses to Semantic Statements. His research into mind/body healing has earned him NeuroEmotional Technique Doctor of the Year Award and the Tree of Life Award for "contributions to the Wellness of the Community at Large." He has received a five-year NIH grant to study the effects of stress reduction on cancer patients.

Dr. Monti is currently writing a scientifically based book for the lay audience on how mind, body and the energetic systems are part of an intricate network that constitutes our physical and mental health.

Andrew B. Newberg, M.D., is an assistant professor in the Department of Radiology and Psychiatry at the Hospital of the University of Pennsylvania and is a staff physician in Nuclear Medicine. He graduated from the University of Pennsylvania School of Medicine in 1993. He did his training in Internal Medicine at the Graduate Hospital in Philadelphia, serving as chief resident in his final year. He has published numerous articles and chapters on the topics of brain function and neuroimaging and the study of religious and mystical experiences. He is co-author of the best-selling book *Why God Won't Go Away: Brain Science and the Biology of Belief* and *The Mystical Mind: Probing the Biology of Belief*, both of which explore the relationship between neuroscience and spiritual experience.

Candace Pert, Ph.D., was a graduate student in her mid-twenties when she discovered the opiate receptor, the cellular bonding site for endorphins, the body's natural painkillers, which she calls our "underlying mechanism for bliss and bonding." This breakthrough presaged a sea change in scientific understanding of human internal communication systems, pointing the way toward the information-based model that is now supplanting the long-dominant structuralist viewpoint.

In the years since, Candace Pert has focused her research on developing nontoxic pharmaceuticals that selectively block receptor sites for the AIDS virus. She has also pursued the "threateningly interdisciplinary" relationship between the nervous and immune systems, developing documentation of a body-wide communication system mediated by peptide molecules and their receptors, which she perceives to be the biochemical basis of emotion and the potential key to many of the most challenging diseases of our time. Her best-selling book, *Molecules of Emotion*, is noteworthy both as an insider's history of the changing scientific paradigm, and as one woman's journey of growth and understanding.

Ramtha (*www.ramtha.com*), one of the great enigmas scientists have studied in the last decade, is a mystic, philosopher, master teacher and hierophant. His partnership with American woman JZ Knight, his channel, still baffles scholars; results of their studies point to a decidedly non-local phenomenon. Using a sophisticated polygraph, noted parapsychologists Ian Wickramasekera and Stanley Krippner of Saybrook Graduate School repeatedly observed that while JZ Knight is channeling Ramtha, the readings of her brain-wave activity shift to delta, and that the lower cerebellum operates her body, which talks, walks, eats, drinks and dances, while Ramtha teaches about the mystery of mind over matter.

Through a coherent system of thought that unifies scientific knowledge with esoteric knowledge of spirit, Ramtha's students study biology, neurophysiology, neurochemistry and quantum physics. Like Bohm, he declares that consciousness is the ground of all being. In his own lifetime, 35,000 years ago, he learned to separate his consciousness from his body, raise its frequency and eventually take it with him. He is one of the few human beings to become an eyewitness to the seen and the unseen.

Dean Radin, Ph.D., is senior scientist at the Institute of Noetic Sciences (IONS) in Petaluma, California. He also serves as adjunct faculty at Sonoma State University and Distinguished Consulting Faculty at Saybrook Graduate School in San Francisco. He earned a BSEE with senior honors, magna cum laude, in electrical engineering from the University of Massachusetts, Amherst, and an M.S. in electrical engineering and Ph.D. in psychology from the University of Illinois, Champaign-Urbana. Early in his career he was a member of the technical staff at AT&T Bell Laboratories and later a principal scientist at GTE Laboratories, where for a decade he was engaged in R&D on a wide variety of advanced telecommunications products and systems.

Radin was elected four times as President of the Parapsychological Association, an affiliate of the American Association for the Advancement of Science.

Dr. Radin is author of the award-winning book *The Conscious Universe* and the forthcoming *Entangled Minds*. He is author or co-author of over 200 journal articles and technical reports.

William A Tiller, Ph.D., is a professor emeritus at Stanford University, a pioneer in psychoenergetics research (currently chairman and chief scientist, The William A. Tiller Foundation for New Science, Payson, Arizona) and author of *Science and Human Transformation: Subtle Energies, Intentionality and Consciousness, Conscious Acts of Creation: The Emergence of a New Physics* and *Some Science Adventures with Real Magic*. He is the recipient of the INTA Humanitarian of the Year Award (1982) and recipient of the 2005 Alyce and Elmer Greene Award for innovation "for his enduring work in psychoenergetics and the role of consciousness in mind-matter reality" from the International Society for the Study of Subtle Energies and Energy Medicine (ISSSEEM).

Jeffrey Satinover, M.S., M.D., is an author, psychiatrist and physicist. His best-selling books, on topics ranging from the history of religion to computational neuroscience and quantum mechanics, have been translated into nine languages. He is currently part of an international multidisciplinary research team at the Condensed Matter Physics Lab at the University of Nice in France investigating complex systems theory with applications to financial markets, climate change, earthquake prediction and social epidemics. He teaches part-time at Princeton University and serves as an advisor to the United States Senate on the trafficking of women and children into sexual slavery. Dr. Satinover has also been the William James Lecturer in Psychology and Religion at Harvard University.

Dr. Satinover holds degrees from MIT, Harvard, the University of Texas and Yale. A graduate of the C. G. Jung Institute of Zurich, he is a former president of the C. G. Jung Foundation of New York. He lives in Weston, Connecticut, with his wife and three daughters, and part time in Saint-Jean Cap-Ferrat, France.

Fred Alan Wolf, Ph.D., works as a physicist, writer and lecturer. His work in quantum physics and consciousness is well known through his popular and scientific writing. He is the author of twelve books, including his latest works: *The Yoga of Time Travel, Dr. Quantum Presents: A Little Book of Big Ideas*, the audio CD series *Dr. Quantum Presents: A User's Guide to Your Universe, Dr. Quantum Presents: Meet the Real Creator: You* and the National Book Award-winning *Taking the Quantum Leap*.

Wolf's inquiring mind has delved into the relationship between human consciousness, psychology, physiology, the mystical and the spiritual. His investigations have taken him from intimate discussions with physicist David Bohm to the magical and mysterious jungles of Peru, from master classes with Nobel Laureate Richard Feynman to shamanic journeys on the high deserts of Mexico, from a significant meeting with Werner Heisenberg to the hot coals of a firewalk. In academia, Dr. Wolf has challenged minds at San Diego State University, the University of Paris, the Hebrew University of Jerusalem, the University of London's Birkbeck College and many other institutions of higher learning.

Dr. Wolf is well known for his simplification of the new physics and the author of *Taking the Quantum Leap*, which, in 1982, was the recipient of the prestigious National Book Award for Science.

Further Suggestions

In addition to all the books, tapes and videos by the incredible minds we interviewed for the film and the book, here is a short list of other favorites we recommend. For a complete list of the books, CDs and videos by our interviewees, as well as a complete recommended reading list, please see our Web site, *www.whatthebleep.com.*

Consciousness:

The Conscious Universe: The Scientific Truth of Psychic Phenomena, Dean Radin, Harper Collins.

Create Your Day, Ramtha, JZK Publishing (DVD).

Entangled Minds, Dean Radin, HarperCollins.

The Science of Mind, Ernest Holmes, Putnam/Tarcher.

Quantum Physics:

Beyond Einstein: The Cosmic Quest for the Theory of the Universe, Michio Kaku, Bantam/Anchor Books.

Beyond the Quantum, Michael Talbot, Macmillan Publishing.

Bridging Science and Spirit: Common Elements in David Bohm's Physics, the Perennial Philosophy and Seth, Norman Friedman, The Woodbridge Group.

The Dancing Wu-Li Masters, Gary Zukav, Bantam.

The Elegant Universe: Superstrings, Hidden Dimensions, and the Quest for the Ultimate Theory, Brian Greene, Random House.

Holographic Universe, Michael Talbot, HarperCollins.

Hyperspace: A Scientific Odyssey Through Parallel Universes, Time Warps and the Tenth Dimension, Michio Kaku, Doubleday/Anchor Books.

The Matter Myth: Dramatic Discoveries That Challenge Our Understanding of Physical Reality, Paul Davies and John Gribbin, Simon and Schuster/Touchstone.

Science and the Akashic Field: An Integral Theory of Everything, Ervin Laszlo, Inner Traditions International.

Science and the Reenchantment of the Cosmos, Ervin Laszlo, Inner Traditions.

Stalking the Wild Pendulum: On the Mechanics of Consciousness, Itzhak Bentov, Destiny Books.

This Strange Quantum World & You, Patricia Topp, Papillon Publishing.

The Tao of Physics: An Exploration of the Parallels Between Modern Physics and Eastern Mysticism, Fritjof Capra, Shambhala.

The Brain:

The God Particle: If the Universe Is the Answer, What Is the Question?, Leon Lederman, Dell.

The Mystical Mind: Probing the Biology of Religious Experience, Eugene G. D'Aquili, M.D., Ph.D., and Andrew B. Newberg, M.D., Fortress Press.

Where God Lives in the Human Brain, Carol Rausch Albright, James B. Ashbrook and Anne Harrington, Sourcebooks.

Where God Lives: The Science of the Paranormal and How Our Brains Are Linked to the Universe, Melvin Morse, M.D., with Paul Perry, Harper San Francisco.

Why God Won't Go Away: Brain Science and the Biology of Belief, Andrew Newberg, M.D., Eugene D'Aquilli, M.D., Ph.D., and Vince Rauss, Ballantine.

Acknowledgments

When we started this film, we thought, "We're just making a movie." What we tapped into was a movement. So many people who have seen our film said, "Finally. I was waiting for something like this to come out!" Well, we were, too; we just realized that we'd probably have to end up doing it ourselves.

We would like to thank everyone for sending their request out in consciousness. The desire for change is palpable now. We are grateful to all of you who supported us in making the film; those of you who supported us at the theaters by sending e-mails and making phone calls to the management sometimes "demanding" to see the film; to all of you who brought your friends and their friends; and those of you in the street teams who pounded the pavements talking about this film. Together we are doing something marvelous.

There are so many people who worked hard behind the scenes to bring the film, and now this book, to the world. Thanks to Gabby, Melissa, Pavel, Cate, Straw, Debbie, Jason, Shelley, John and David for their never-ending support—hours upon hours of working the phones, the Internet and whatever else it took to bring the film to the masses. Also everyone at Samuel Goldwyn, IDP and Twentieth Century Fox Home Video for believing in the film, and Richard Gaurdian for taking it worldwide.

Thank you to everyone at our publisher, HCI, and especially to Amy Hughes, our editor; Larissa Hise Henoch, who created most of the artwork; and Bret Witter, who oversaw the entire project.

Thanks to our families and friends for allowing us the space and time it took to bring our dreams into reality: Gordie and Elorathea, for personal inspiration (Betsy); Mary Lou, for all the years of love and support (Mark); and for keeping me smiling and laughing through it all: Walter, Viva, Rama and K. T. Elliott (Will).

To Ramtha & JZ Knight, without whom none of this would have been possible. To all of our scientists and interviewees—Fred Alan Wolf, Dean Radin, Lynne McTaggart, Dan Monti, David Albert, Jeffrey Satinover, Stuart Hameroff, Amit Goswami, Bill Tiller, Miceal Ledwith, Joe Dispenza, Candace Pert, Andrew Newberg, Ervin Laszlo and John Hagelin—thank you for your brilliant minds and your willingness to go where most men and woman are afraid to.

The new frontier is not a country or a planet. It is mind. These ideas are about the future of humanity.

Will Betsy Mark